Assistive Technology Design for Intelligence Augmentation

Synthesis Lectures on Assistive, Rehabilitative, and Health-Preserving Technologies

Editor

Ron Baecker, *University of Toronto*

Advances in medicine allow us to live longer, despite the assaults on our bodies from war, environmental damage, and natural disasters. The result is that many of us survive for years or decades with increasing difficulties in tasks such as seeing, hearing, moving, planning, remembering, and communicating.

This series provides current state-of-the-art overviews of key topics in the burgeoning field of assistive technologies. We take a broad view of this field, giving attention not only to prosthetics that compensate for impaired capabilities, but to methods for rehabilitating or restoring function, as well as protective interventions that enable individuals to be healthy for longer periods of time throughout the lifespan. Our emphasis is in the role of information and communications technologies in prosthetics, rehabilitation, and disease prevention.

Assistive Technology Design for Intelligence Augmentation
Stefan Carmien

ISBN: 978-3-031-00473-5   print
ISBN: 978-3-031-01601-1   ebook

DOI 10.1007/978-3-031-01601-1

A Publication in the Springer series
SYNTHESIS LECTURES ON ASSISTIVE, REHABILITATIVE, AND HEALTH-PRESERVING TECHNOLOGIES, #9
Series Editors: Ronald M. Baecker, University of Toronto

Series ISSN 2162-7258 Print   2162-7266 Electronic

Assistive Technology Design for Intelligence Augmentation

Stefan Carmien

The Tecnalia Foundation

SYNTHESIS LECTURES ON ASSISTIVE, REHABILITATIVE, AND HEALTH-PRESERVING TECHNOLOGIES #9

ABSTRACT

Assistive Technology Design for Intelligence Augmentation presents a series of frameworks, perspectives, and design guidelines drawn from disciplines spanning urban design, artificial intelligence, sociology, and new forms of collaborative work, as well as the author's experience in designing systems for people with cognitive disabilities. Many of the topics explored came from the author's graduate studies at the Center for LifeLong Learning and Design, part of the Department of Computer Science and the Institute of Cognitive Science at the University of Colorado, Boulder. The members of the Center for LifeLong Learning and Design came from a wide range of design perspectives including computer science, molecular biology, journalism, architecture, assistive technology (AT), urban design, sociology, and psychology.

The main emphasis of this book is to provide leverage for understanding the problems that the AT designer faces rather than facilitating the design process itself. Looking at the designer's task with these lenses often changes the nature of the problem to be solved.

The main body of this book consists of a series of short chapters describing a particular approach, its applicability and relevance to design for intelligence augmentation in complex computationally supported systems, and examples in research and the marketplace. The final part of the book consists of listing source documents for each of the topics and a reading list for further exploration.

This book provides an introduction to perspectives and frameworks that are not commonly taught in presentations of AT design which may also provide valuable design insights to general human–computer interaction and computer-supported cooperative work researchers and practitioners.

KEYWORDS

intelligence augmentation, assistive technology, design frameworks, cross-discipline design, task support for persons with cognitive disabilities, computer-supported work

Contents

Figure Credits

Figure 1.1: From: L3D. (2006). "Center for LifeLong Learning and Design, University of Colorado, Boulder." Copyright © 2006 Center for Lifelong Learning and Design. Used with permission.

Figure 1.2: From: Coleman. Coleman Institute for Cognitive Disabilities website. Copyright © 2004 Coleman Institute for Cognitive Disabilities. Used with permission.

Figure 2.1: Author's picture, 2015, Carrefour Saint Jean De Luz.

Figure 2.2: From: Carmien, S. P. and G. Fischer (2008). Design, adoption, and assessment of a socio-technical environment supporting independence for people with cognitive disabilities. *Proceedings of the Twenty-Sixth Annual SIGCHI Conference on Human Factors in Computing Systems* (Florence, Italy, April 05–10, 2008). CHI '08 ACM, New York, pp. 597–606. Copyright © 2008 ACM. Used with permission.

Figure 2.3: From: Carmien, S. P. and G. Fischer (2008). Design, adoption, and assessment of a socio-technical environment supporting independence for people with cognitive disabilities. *Proceedings of the Twenty-Sixth Annual SIGCHI Conference on Human Factors in Computing Systems* (Florence, Italy, April 05–10, 2008). CHI '08 ACM, New York, pp. 597-606. Copyright © 2008 ACM. Used with permission.

Figure 2.4: From: Carmien, S. (2011). Socio-technical environments and assistive technology. In *Socio-technical Networks: Science and Engineering Design.* F. Hu, A. Mostashari, J. Xie, Eds. Boca Raton, FL, Taylor & Francis LLC, CRC Press: 167–180. Copyright © 2011 Taylor & Francis Group, LLC. Used with permission.

Figure 2.5: From: Assistant Project (2015). Aiding SuStainable Independent Senior TrAvellers to Navigate in Towns. EU, AAL. Used with permission.

Figure 2.6: From: Assistant Project (2015). Aiding SuStainable Independent Senior TrAvellers to Navigate in Towns. EU, AAL. Used with permission.

Figure 2.7: From: Assistant Project (2015). Aiding SuStainable Independent Senior TrAvellers to Navigate in Towns. EU, AAL. Used with permission.

Figure 2.8: From: Assistant Project (2015). Aiding SuStainable Independent Senior TrAvellers to Navigate in Towns. EU, AAL. Used with permission

List of Abbreviations

AAC	Augmentative and Alternative Communication
ADA	Americans with Disabilities Act
ADL	Activities of Daily Living: basic activities including eating and dressing
AI	Artificial Intelligence
ANOVA	Analysis of Variance
ASSISTANT	An AAL EC project: Aiding SuStainable Independent Senior TrAvellers to Navigate in Towns
AT	Assistive Technology
AT/IA	Assistive Technology using Intelligence Augmentation
CLever	The CognitiveLEVERs project in L3D, University of Colorado
COACH	The Cognitive Orthosis for Assisting with aCtivites in the Home project from Intelligent Assistive Technology and Systems Lab (IATSL) in the University of Toronto
DC	Distributed Cognition
DFA	Design for All
DM	Device Model
DRD	Digital Resource Description
EU4ALL	European Union for All an EC FP7 accessibility project
GPRS	General Packet Radio Service
GPS	Global Positioning System
GUI	Graphical User Interface
HAPTIMAP	Haptic, Audio, and Visual Interfaces for Maps and Location Based Services, a FP7 EC project
HCI	Human-Computer Interaction
HIPAA	The US Health Insurance Portability and Accountability Act
IA	Intelligence Augmentation
IADL	Independent Activities of Daily Living: more complex daily activities
ICF	International Classification of Functioning, Disability, and Health of the WHO

L3D	The Center for LifeLong Learning and Design of the Department of Computer Science and the Institute of Cognitive Science at the University of Colorado
MAPS	MAPS Memory Aiding Prompting System
MAPS-DE	Script editor for the MAPS Memory Aiding Prompting System
MAPS-PR	The user interface for the MAPS Memory Aiding Prompting System
PC	Personal Computer
PDA	Personal Digital Assistant (a precursor to the Smartphone)
RAID	Redundant Array of Independent Disks
SMS	Short Message Service
STE	Socio-Technical Environment
STS	Socio-Technical System
TYPE-1 ERROR	Classifying an event as occurring when it has not
TYPE-2 ERROR	Classifying an event as not occurring when it has
UI	User Interface
UM	User Model
WAIS-IQ	Wechsler Adult Intelligence Scale

Acknowledgments

The Tecnalia Foundation supported the writing of this book along with the AAL-funded ASSISTANT project (inspired by Jim Sullivan's work in his Mobility for All Clever project). Many of these ideas were incubated in The Center for LifeLong Learning and Design (L3D) at the University of Colorado under the leadership of Gerhard Fischer and the enriching association of the members and visitors of L3D, most notably Anja Kintsch, Andrew Gorman, Hal Eden, and Ernesto Arias. My dissertation committee extended my perspective, especially Clayton Lewis, Leysia Palen, and Cathy Bodine and The Coleman Institute for Cognitive Disabilities funded much of my dissertation work. Finally, I would like to thank the young adults with cognitive disabilities, elders, parents, and professionals I have had the good fortune to work with over the years.

Preface

Most of the ideas in this book came to me while I was a student of Gerhard Fischer over the years in his Centre for LifeLong Learning and Design. I had the enormous good fortune to have him as a dissertation advisor and spent five years in the lab soaking up our weekly meetings and being exposed to the widely disparate experts that flowed through—sometimes for the weekly gatherings to give a talk, sometimes visiting and working for months. Beyond this, I brought insights from the last ten years of research in Europe at the Fraunhofer Institute at Bonn under Dr. Carlos Velasco and currently at Tecnalia foundation in San Sebastián Spain. Prior to my academic and research career, I spent 12 years as a production and inventory control manager in manufacturing companies, and prior to that I was a cabinetmaker making custom furniture. I hope that the sum of experiences and teachers I brought to this book can be part of a useful foundation for "pushing back the frontiers of science" as my doktorvater, Gerhard Fischer, would exhort me to do.

Part 1

CHAPTER 1

Introduction

I am in the field of designing and evaluating technology for people with disabilities, especially for those with cognitive disabilities. Doing this, I have often drawn, particularly in early stages of a project, on the ideas that were "in the air" while in graduate school, in the Centre for Lifelong Learning and Design (L³D, 2006) at the University of Colorado at Boulder (Figure 1.1). This is because the nature of the problem and the target population do not lend themselves to easily described requirements and quantifiable results. Let's go back over this somewhat cryptic sentence. In this domain the problem often is the mismatch between the cognitive skills available and the task's cognitive requirements, for instance performing a job that is just a bit too complex to remember all the steps (Lancioni et al., 2000). This is very different than magnifying a screen or reading aloud the text on a screen to compensate for visual deficiencies. For example, the Americans for Disabilities Act (ADA, 1990) has a very nice quantifiable list of requirements (ADA, New England ADA Center, 2015) for accessibility of buildings, mostly for mobility issues (e.g., wheelchair access), that make it easy to see if the building's accessibility requirements are met. There is no such simple set of requirements for accessible and useful cognitive technology, although there are standards committees (ISO, 2015; RESNA, 2015) working on this thorny issue. Additionally, the environment of the job may change as well as the specifics of the task at hand (Suchman, 1987), whereas screens and users almost always remain the same relation. As for the population, mitigating missing intellectual deficiencies is both complicated and unique to each person, due to individual cognitive variations (Cole, 2013) and co-existing illnesses and pathologies (Mc Sharry, 2014).

1.1 LAY OF THE LAND

Before continuing, we need to unpack several concepts: (1) what is meant here by "peple with cognitive disabilities" and what are their unique needs; and (2) what problems we are trying to solve and finally there will be several examples of the types of technology targeted.

The end-users for these technologies can be roughly described as people with cognitive disabilities, but this broad category can be misleading. More precisely, people who need these support systems could be functionally described as missing the cognitive functions (congenitally, by (de)acquisition, or by gradual cognitive decline) that make them capable of determining and taking the actions needed to live a high quality of life, because of deficiencies in memory, executive function, or the ability to derive accurate and appropriate decisions based on the events in ordinary life. This is a functional definition (Scherer, 2011), so it does not particularly matter what the etiology of impaired cognition is, i.e., the type of congenital problem in development that is the basis of the functional deficiency (with the exception of, to some extent, the peculiarities of Autism syndrome—see Section 5.1). Moreover, nor does the manner of losing cognitive acuity and mnemonic ability—such as acquired traumatic brain injury or diseases of aging like Alzheimer's (in the early stages) or mild cognitive impairment—make for different sets of requirements. In this world, every cognitive impairment of sufficient morbidity becomes a "universe of one" to be designed for and the cause often bears as little importance as the geographic location. Now the population is functionally defined, as will become clearer in the discussion of the International Classification of Functioning, Disability, and Health (ICF) (World Health Organization, 2001) in the Section 5.1.

To what kinds of support systems are the concepts' described here applicable? Just as every artificial intelligence (AI) problem can be described as a classification problem (or a set of such problems (Russell and Norvig, 2009)), the applications here are all versions of task support. Task support spans shopping, cooking, navigation (by bus or walking), planning activities, taking medication, washing hands, doing a job, deciding what to cook, using money or a computer, deciding where to go and how to get there, and what to do when lost or off track. All and anything to do with day-to-day tasks, from very simple to complex sets of tasks. That's a lot of ground to cover. Some of the examples are very focused, such as helping elders with cognitive impairment properly wash their hands (which turns out to be a very hard task) (Mihailidis et al., 2008) to traveling across the county to visit your sister (Assistant Project, 2015). What we are *not* talking about are specific sensory issues associated with some forms of intellectual disability, such as dyslexia or dyslisia.

One of the interesting side effects of working in this area is that, for almost any application discussed, someone will say that they could use the system themselves from time to time. The now-familiar curb cut effect (Carmien et al., 2005), when it comes into contact with context-caused functional cognitive disabilities leads to many opportunities like using a public transportation navigation system to support tourists who are new to the area and non-native speakers of the local language. However, care must be taken to avoid creeping featurism, ending up with a system difficult to use by the original target population.

1.2   ASSISTIVE TECHNOLOGY (AT): ADOPTION AND ABANDONMENT

Device rejection is the fate of a large percentage of purchased assistive technology (King, 1999; 2001). Caregivers report that difficulties in configuring and modifying configurations in assistive technology often leads to abandonment[1] (Kintsch and dePaula, 2002), an especially poignant fate considering that these types of systems may cost thousands of dollars. While assistive devices can

---

[1] There is another kind of abandonment, which is not using the system or device because the need no longer exists. This "good" abandonment of AT is not in the purview of the current study.

have a profound effect on life, such devices have a high abandonment rate, ranging from 8% for life-saving devices to 75% for hearing aides. In fact, about one-third of all assistive devices are abandoned (Scherer, 1996; Scherer and Galvin, 1996). While there are no studies examining the abandonment rate across all types of assistive devices (Kintsch and dePaula, 2002), some experts estimate that as much as 70% (Martin and McCormack, 1999; Reimer-Reiss, 2000) of all such devices and systems are purchased and not used over the long run, particularly those designed as a cognitive orthotic (LoPresti et al., 2004). Other causes for abandonment have many dimensions; a study by Phillips and Zhao reported that a "change in needs of the user" showed the strongest association with abandonment (Phillips and Zhao, 1993). Thus, those devices that cannot accommodate the changing requirements of the users were highly likely to be abandoned. It then follows logically (and is confirmed by interviews with several AT experts (Kintsch, 2002; Bodine, 2003)) that an obstacle to device retention is difficulty in reconfiguring the device. A survey of abandonment causes lists "changes in consumer functional abilities or activities" as a critical component of AT abandonment (Galvin and Donnell, 2002). A study by Galvin and Scherer states that one of the major causes for AT mismatch (and thus abandonment) is the myth that "a user's assistive technology requirements needs to be assessed just once" (Scherer and Galvin, 1996); on-going re-assessment and adjustment to changing needs is the appropriate response. A source for research on the other dimensions of AT abandonment, and the development of outcome metrics to evaluated adoption success, is the ATOMS project at the University of Milwaukee (Rehabilitation Research Design & Disability (R2D2) Center, 2006). The types of AT design to support task completion, decision making, and navigation are specifically most at risk for non-adoption or abandonment. It is these kinds of systems this book is aimed at helping the designer succeed with.

Successful AT design for this population must support the interface requirements for users with cognitive impairments as well as view configuration and other caregiver tasks as different, yet equally, important requirements for a second user interface (Cole, 1997). One proven approach applies techniques such as task-oriented design (Lewis and Rieman, 1993) to mitigate technology abandonment problems. Research (Fischer, 2001b) and interviews (Kintsch, 2002) have demonstrated that complex, multifunctional systems are the most vulnerable to abandonment due to the complexity of the many possible functions.

It is very encouraging to see an increasing educational focus on the design of assistive technology in engineering schools across America. One of the useful marks of many of these courses is the involvement of hands-on professionals, often in the form of special education teachers or vocational rehabilitation specialists, insuring that not only is the technology built right (or in software engineering terms validated as correct (IEEE, 1990)) but that it is the right form of technology for the user/problem (in software engineering verified as the right system). This is a big step forward from what was more common in the 1990's and 2000's when there seemed to be a lot of designs coming out of naive but inspired systems that seemed more an exploration of what could be done rather than what was needed. But, beyond involving members from various stakeholder groups (see Section 4.5 in Part 2), there are other ways of looking at needs and implementing solutions, from other disabilities and from more deeply looking at assistive technology (AT) design itself. Exploring some of these ideas are what this book is about. L3D's approach to design, by looking deeply at the context and pattern of problems as problems has proved remarkably useful in working with these complex, dynamic, and often idiosyncratic challenges.

1.3   ASSISTIVE TECHNOLOGY AND THE TOOLKIT

The ideas here comprise what I think of as my AT design toolkit—a set of conceptual guides and levers to support and contextualize AT and Design for All[2] (DfA (Stephanidis and ASavidis, 2001; Center for Universal Design, 2011)) approaches. The notion of a toolkit comes from Tammy Summers (1995):

> I refer to these software collections as "high-tech toolbelts" because each designer assembles
> her personal collection just as a carpenter assembles a collection of hammers, screwdrivers,
> tape measures, etc. into a personal toolbelt.

While this refers to the strengths and problems involved in using multiple tools to create representations of problems, the notion of a "toolbelt" (or in this case a "toolkit") is, I think, a good one. We all bring toolkits to design problems, and all problems that are not simple assembly of pre-created parts are design problems. For some of us, this is simply the set of skills that have worked for us in the past—perhaps they have been taught to us in a formal fashion (i.e., schooling or apprenticeship), or perhaps they are part of the cultural zeitgeist we were raised with. My approach is to exteriorize the choice of the tools[3] and lenses to approach each design project. By exteriorize I mean that after first carefully and un-biasedly looking at the problem or situation, one steps back and considers what might be applied to make the process tractable or even just what this is similar to in previous experience.

So what do you look at? The simple answer is the end-user and what they want to, or must, do. To think about end-users you end up thinking about the context of use and the larger goal of use. The ideas and tools here are intimately tied to this investigation.

---

[2]   AT and DFA refer to two overlapping/complementary approaches to accessibility design. AT technologies make adaptations that allow users to do something that they otherwise would be unable to. DFA is a design movement to guide designers to create systems and artifacts that are useable by as many different types of people (e.g. with disabilities) as possible. While the ultimate goal of DFA is to make all human technologies accessible by all people of varying abilities, the realties of the distribution of abilities (i.e. some disabilities are so unusual that designing a system for everyone is either too hard or expensive to do) make AT one end of the continuum that will continue to be needed.

[3]   By "tools" I do not necessary mean tools in the sense of CASE or IDE tools, although these are included; tools here mean any idea, framework, or theory that extends our ability to understand or create systems.

Where did the toolkit come from? The majority of ideas here came from time in my graduate school at the University of Colorado in the Centre for Lifelong Learning and Design. So what does the L³D approach have to say about this? The general approach can be summarized with this quote (L³D, 2006):

> The Centre for Lifelong Learning and Design is part of the Department of Computer Science and the Institute of Cognitive Science at the University of Colorado at Boulder. The mission of the center is to establish, both by theoretical work and by building prototype systems, the scientific foundations for the construction of intelligent systems that serve as amplifiers of human capabilities.

Amplifiers of human capabilities… This is supported by five basic dimensions brought together from widely different research cultures (Fischer, 2002; 2003):

- artificial intelligence (AI) → intelligence augmentation (IA);

- instructionist learning → constructionist learning;

- individual focus → social contexts;

- things that think → things that make us smart; and

- what computers can do → human and computer synergies.

This book discusses, in some depth the dimensions that particularly relate to the task of designing AT. L³D members are drawn from computer science, education, psychology, electrical engineering, architecture/urban planning, microbiology, and sociology/anthropology. L³D has two parent organizations: the Department of Computer Science and the Institute of Cognitive Science. L³D interacts with other academic units at CU Boulder (College of Architecture and Planning), K-12 schools, community groups, government laboratories (NCAR/UCAR), and industrial partners (BEA, Siemens, IBM, Apple, PFU, SRA).

Not all of the elements in this book come specifically from L³D; some come from my experience in developing ATs in the lab while being supported by the Coleman Institute for Cognitive Disabilities (Figure 1.2). The Coleman Institute formed the bridge between L³D and ATs. The Coleman Institute became interested in the work that L³D was producing, particularly its unique approach, and I was fortunate enough to be supported by them during my Ph.D. studies. L³D formed a group of developers and researches that called themselves CLever (Cognitive LEVERs) (CLever, 2004) , which produced several projects, many papers, and two dissertations. CLever pulled together interest in AT, our expertise in design and cognitive science, and, most importantly, external input from domain experts and the L³D community. The domain experts (one half-time staffer and any number of invited speakers at the meetings) brought in expertise in AT design and many, many years of experience in design and computer-assisted work in other contexts. They provided tremendous leverage in avoiding naïve mistakes (see Section 5.1 as well as Section 5.2). The

larger L³D community provided us with the sort of "out of the box" questions and criticism that supported effective novelty in our designs (see Section 4.5).

Figure 1.1: L³D logo. From L³D (2006).

Figure 1.2: Coleman Institute logo. From Coleman (2004).

1.4   ELEMENTS OF THE TOOLKIT

The following sections will examine, in a framework, the 20 perspectives that I have chosen for this book (Table 1.1). Each topic will begin with an explanation of the concept, and where it came from and how it was initially used. Then, the bridge from the contact to intelligence augmentation and AT is presented. Finally, the section presents several examples of the concept in use or in current research. The documents seemed to fall naturally into four categories: Fundamentals, Models, Technique, and Things to Avoid. The table below shows the topics and categories.

Table 1.1: Categories and concepts	
Category	**Element**
Fundamentals	Artificial intelligence (AI)/Intelligence augmentation (IA)
	Design for failure
	Distributed cognition
	Scaffolding
	Situated action/cognition
	Socio-technical environments
	Universe of one
	Wicked problems
Model	Importance of representation
	Tools for living and tools for learning
	Dyads
Technique	Plans and action
	Low-hanging fruit
	MetaDesign
	Personalization
	Symmetry of ignorance
Things to Avoid	Diagnosis and functionality
	I have a theory/cousin
	Islands of ability

The collections don't follow any meta categorization plan, and are not necessarily sequential, however with the exception of things to avoid the collections roughly fall into what can be called the ground of design, the development or path of the process, and the implementation or fruition of the design. The first category, Fundamentals, presents tools that can be used in general approach of the problem, for instance thinking about how to leverage the end-user's abilities rather than just produce results—which is one of the differences between AI and IA. Modeling items present ways to think about how to present the problem and also the kinds of solutions that might result. The Technique is just that, specific frameworks that can be used to approach these high function in AT/DfA problems. Finally, Things to Avoid is a catch-all category that came out of my initial work in AT and the shortcuts of domain experts, such as the special education technology expert we had the good fortune to have on part-time in the lab during the CLever (CLever, 2004) project. There are so many blind alleys in developing this sort of technology and having a domain expert with years of experience available to point out dead ends and acting as a proxy user in end-user design process is invaluable. If you were starting out in this field I would recommend finding a partner such as Anja Kintsch who guided us (see Section 4.5).

Part 2 will discuss each of these topics. For each one there is an initial definition of the concept, then a more detailed discussion with respect to assistive technology design for intelligence augmentation, a list of canonical papers and discussion of the source of the concept, and finally examples of the concept in existing systems, with an emphasis on how its design was influenced by the concept. Most of the examples in Part 2 are only presented on the basis of my familiarity with the systems, a result of time spent talking about them with their creators. The other criteria of systems discussed follows my personal commitment to working on projects that promise to mitigate the digital divide and to make the cost of a commercial version reasonable for those of us not so financially well off.

Part 3 lists a set of publications about each concept, in the same order as Part 2. The publications go into the concepts in more detail, some of them relating to the source of the concept, some of them about specific implementations and forks of the concept in different domains.

1.5   HOW TO USE THIS BOOK

While I certainty can't compare the depth of insights and broad and profound applicability in this book to the "Gang of Four"'s seminal *Design Patterns* (Gamma et al., 1995) or to Christopher Alexander's *A Pattern Language* (Alexander et al., 1977), the approach is similar. Depending on your interest and needs, this can be used as a broad overview of the landscape of designing and implementing these types of systems, as an introduction to the source of the concepts, and of course as a guide to using them in the design and construct in this particularly interesting area of technology.

Some of the descriptions that follow only expose the concept and point to relevant publications; concrete examples are only briefly discussed as it is difficult to extract specific design components that illustrate a broad design approach (Situated Action/Cognition, Wicked Problems). Other descriptions have detailed examples, in one case illustrating the complexity and detail of the implementation but not intended as an algorithm to copy (Personalization), and in another a very implementable set of steps to integrate this approach into your system (Design for Failure). For all but one concept Part 3 lists publications that were seminal to the concept and in many cases others that illustrate various ways in which the concept is implemented.

I have intentionally left out some of the more advanced platforms and approaches that will become more available and mainstream in the future. These include the Internet of Things⁴ (IoT), intelligent agents, and "emotion"-based robotics. This book is meant to help the development of Assistive Technology using Intelligence Augmentation (AT/IA) now, and in the very near future, not to propose or discuss the cutting edge work that will make production of IA systems much more effective and easier to use and design in the not-so-near future. However, the topics to follow will be useful and important no matter what platform or infrastructure you design with.

---

4   This assertion is made in the sense of the IoT as being integral to an AT system. This excludes GPS as an IoT component and recently developed IoT systems that straddle SmartHomes, requiring high levels of infrastructure spending. Similarly, there are many, many indoor navigation systems that require putting sensors all over the building to succeed. What has interested me is the creation of intelligent AT that does not exclude the majority of potential users due to expense or the need for everyone to adopt a standard (especially in the current world of many competing, possible, standards). I have no doubt that the interconnected world of IoT will, in the near future, produce affordable solutions based on an existing infrastructure and opportunistically available information without proprietary shackles. But I cannot yet discuss rules of thumb or theoretical frameworks that are unique to IoT systems. I have, however, tried to present the concepts in Part 2 so that they may be applicable in IoT-based systems.

Part   2

THE CONCEPTS

This part of the book discusses each of the topics presented at the end of the last section. I have tried to give you enough material in each topic to allow you to start using it. For each one there is:

1. an initial definition of the concept;

2. a more detailed discussion with respect to Assistive Technology Design for Intelligence Augmentation;

3. a canonical paper(s) and a discussion of the source of the concept;

4. example(s) of the concept in existing systems, with an emphasis on how its design was influenced by the concept.

Finally, in some cases I have added a conclusion to tie it all up.

The topics are grouped as in Table 1.1, into Fundamentals, Models, Technique, and Things to Avoid.

CHAPTER 2

Fundamentals

This first section of concepts, Fundamentals, represents a set of perspectives that are useful when considered as frameworks for thinking about how to investigate and plan a design for the problem you are considering. The big difference between this and the following sections is that, in my experience, each of these fundamentals are necessary to be considered in any AT/IA project, whereas the concepts in the sections on Models and Technique may be only applicable to certain kinds of projects. In this fundamental category are the following topics:

- artificial intelligence (AI)/intelligence augmentation (IA);

- design for failure;

- distributed cognition;

- scaffolding;

- situated action/cognition;

- socio-technical environments;

- universe of one; and

- wicked problems.

2.1 ARTIFICIAL INTELLIGENCE (AI) → INTELLIGENCE AUGMENTATION (IA)

Short Definition: Instead of using technology and machine learning to replace cognitive behavior (artificial intelligence), use these tools to leverage existing cognitive abilities to achieve the same ends. It is arguable that the history and key developments of HCI are based on this approach. Mice, desktops, cut, and paste are all based on a vision of IA (see the first chapters of Markoff's remarkable book on the early development of personal computers (Markoff, 2005))

   Longer Description: Traditionally, supporting end-users with AI meant that the problem that they were facing, but could not accomplish (e.g., deciding what choice to take, what to do next, how to get a desired result), would be solved for them and the solution presented to the user. The IA approach instead provides tools and information to leverage existing ability of the user to solve the problem. Here is an example. In the near future, smart kitchens will be able to know the contents of the family's pantry and refrigerator. From this, the AI approach might be to generate a shopping list (or better still to order for delivery) based on past stock levels; the IA approach could instead suggest what items to purchase based on possible meals that could be made with the current food in the home, and also make suggestions about replenishing the commonly used stock in the kitchen, and based on the responses to the suggestions it alters the expectations about what the family needs to buy on an on-going basis. A less futuristic example is spelling correctors: the AI model simply replaces the "wrong" text with what it assumes the writer meant, and the IA version presents a choice of words, including adding the "wrong" spelling to the list of correctly spelled words.

There are several benefits to splitting the act of cognition rather than making the decision for the end-user.

- Let each part of the cognitive process do what it does best. Computational machines count, enumerate, estimate probabilities, and control motors better than humans; humans know their desires, visually identify things (in general and for now), understand the context of the task in a nuanced way, and feel most comfortable with a sense of control. By splitting the task and fitting the appropriate parts into the best place, the task becomes more tractable and the user is more satisfied with the results.

- One of the results of supporting people by leveraging the end-user's abilities to produce the cognitive act (e.g., making a decision), rather than replacing cognitive acts of the end-user with a pre-calculated answer or decision, is that you give the human the feeling of being active in the process. This supports a sense of personal dignity and autonomy. By putting them in a active relationship with the cognitive process, they are committed and motivated to adopt the system better than just being a passive recipient of instructions. Moreover, it facilitates context-sensitive cognition.

There are two interesting themes in changing the design focus from using AI to using IA to support tasks or decision activities. First is the Replacement → Empowerment axis. When computational support consists of replacement of the cognitive act with pushing a complete solution to the user several things can happen: the answer is not MY answer, the answer is wrong, and the answer may return the desired goal but the intermediate steps are not workable or desirable. The difference between replacement and empowerment can be the difference between adoption and abandonment. When the user feels ownership and participation in the cognitive process they are more motivated to use the result. Next is the Emulate → Complement axis. What is being pointed out there is that in the case of AI the machine/system is not just doing what it is best at but it is also being forced to emulate the human cognitive process (if only at the "goodness of solution" utility function level). However, if the solution process involves both the end-user and the AI system the parts complement each other and thus are more effective. Of course this assumes, like most economic theories, that users are rational (i.e., homo economicus) and choose optimally, which sometimes is not the case at all, something the designer needs to keep in mind.

An excellent example of the efficiency of the IA approach (but not necessarily the empowerment aspects of IA) is the system for self-pricing produce provided in some Carrefour supermarkets in France (Figure 2.1). Typically, in European supermarkets the shopper is expected to self-price their selected produce. This usually involves putting the produce in a plastic bag and looking at a number on the description/price label next to the produce bin, then taking the bag to a pricing scale and pushing the number associated with the type of produce on a screen. The scale then emits an adhesive sticker with a barcode, description, and price of the bag of produce. The way that these stores implement self-pricing is that you bag your purchase, and place it on the scale, repeating this for each kind of produce they want. Based on information gathered from a camera and the scale itself (Bronnec, 2015) (e.g., the characteristics of the produce), the screen produces four possible identifications of the produce. If it is in the list you touch that picture (see Figure 2.1), if not, the scale allows you to input the name of the item on a touch keyboard on the screen. For me it is right about 90% of the time.[5] This eliminates the cognitive load of keeping the produce ID number in your head while you are walking to the scale (which can be quite a problem with those of us with short-term memory issues), as well as not forcing you to go to the scale after each selection, you could choose all your produce then do all the pricing.

One of the marks of IA design is that it leverages existing knowledge and metaphors commonly known by end-users. This requires designers to be familiar with the target end-users, an existing requirement for good design, but in this case not understanding the end user may lead to failure in use. This may be particularly important for new adopters or those on the wrong side of the digital divide; what are assumed to be helpful shortcuts become daunting fences.

---

[5]   Interestingly, on the top of the machine is a sign alerting shoppers that there is a video camera on. This is an excellent example of what happens when you do not do due-diligence in shareholder analysis. I am guessing that the application was complete and tested well enough to deploy when a legal expert from another part of the company forced them to add this laminated cardboard sign to comply with EU privacy laws.

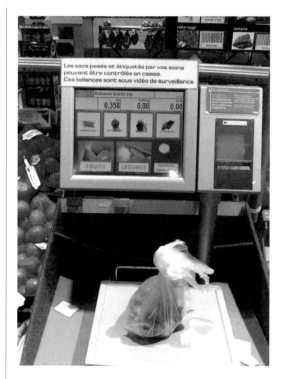

Figure 2.1: Carrefour vegetable pricing system. Author's picture, 2015, Carrefour Saint Jean De Luz.

2.1.1   CANONICAL PAPER

There are many streams that lead into this topic (see the extended references in Part 3). However, this one source stands out for both an early vision of the direction, and for its wide citation:

> Engelbart, D. *Augmenting Human Intellect: A Conceptual Framework* (Engelbart, 1962).

2.1.2   AT EXAMPLES

The first AT example is the MAPS (Memory Aiding Prompting System) (Carmien, 2006b), a research system to support adults with cognitive disabilities performing activities or tasks that are too difficult for them to do unaided. MAPS ran on a PDA (in this case a HP ipaq) platform to present verbal and pictorial prompts in sequence. Furthermore, this set of prompts comprised a script that guided the user through completing a task. The PDA provided error correction functionality via dynamic, situated scripting and "panic button" functionality (using wireless connectivity). A PC-based application provided tools for script creation, modification, and sharing with other users via a web-based repository of scripts. A web script repository has numerous tested script templates assisting the design of new personalized successful scripts.

MAPS research has explored independence specifically in the following contexts: (1) to extend the ability to choose and do as many activities of daily living as possible; (2) to be employed, but without the constant or frequent support and supervision of a professional job coach; and (3) to prepare meals and shop for weekly groceries. Independence is not at odds with socialization; it is the foundation of inclusion and engagement in society.

The MAPS Environment

MAPS (Carmien, 2007) consists of two major subsystems that share the same fundamental structure but present different affordances for the two sets of users: (1) MAPS-DE, for caregivers, employs web-based script and template repositories that allow content to be created and shared by caregivers of different abilities and experiences; and (2) MAPS-PR, for clients, provides external scripts that reduce the cognitive demands for the clients by changing the task.

The MAPS-Design-Environment (MAPS-DE)

The scripts needed to effectively support users are specific for particular tasks, creating the requirement that the people who know about the clients and the tasks (i.e., the local caregivers rather than a technologist far removed from the action) must be able to develop scripts. Caregivers generally have no specific professional technology training nor are they interested in becoming computer programmers. This creates the need for design environments with extensive end-user support to allow caregivers to create, store, and share scripts (Fischer and Giaccardi, 2006). Figure 2.2 shows MAPS-DE for creating complex multimodal prompting sequences. The prototype allows sound, pictures, and video to be assembled by using a filmstrip-based scripting metaphor.

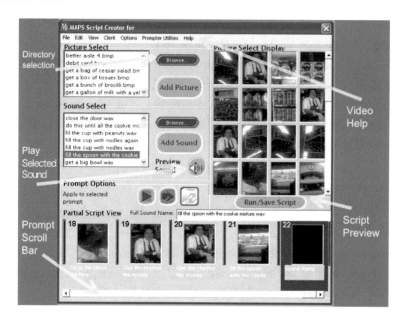

Figure 2.2: The MAPS-design-dnvironment for creating scripts. From Carmien and Fischer (2008).

MAPS-DE supports a multi-script version that allows caregivers to present the looping and forking behavior that is critical for numerous task-support situations. MAPS-DE was implemented on a Microsoft OS (Windows 2000 or XP) platform connecting to and supporting PDAs that run the WIN-Compact Edition (WIN-CE) operating system.

The MAPS-Prompter (MAPS-PR)

MAPS-PR presents to clients the multimedia scripts that support the task to be accomplished. Its function is to display the prompt and its accompanying verbal instruction. MAPS-PR has a few simple controls (see Figure 2.3): (1) the touch screen advances the script forward one prompt and (2) the four hardware buttons on the bottom, which are mapped to: (i) back up one prompt, (ii) replay the verbal prompt, (iii) advance one prompt, and (iv) activate panic/help status. The mapping of the buttons to functions is configurable to account for the needs of individual users and tasks.

The current platform for the MAPS-PR is an IPAQ_3850. The system was implemented for any machine that runs the WIN-CE operating system. MAPS-PR has been installed on units that have cell phone and GPRS functionality. The prompter software was originally written in embedded Visual Basic, and then ported to the faster and more flexible C# .net environment. The prompter software supports single-task or multi-task support.

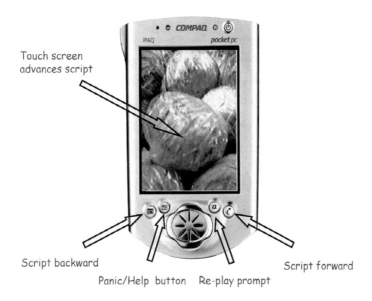

Touch screen
advances script

Script backward

Panic/Help button   Re-play prompt

Script forward

Figure 2.3: The MAPS Prompter (MAPS-PR). From Carmien and Fischer (2008).

More to the point, the MAPS system was tested in training a 19-yr old with developmental disabilities to sort clothes in a very specific way (i.e., by type, then by size, then by color (in a very definite order)). She was working with a job coach in the traditional way and switched to using MAPS to support doing the task. In this case the intent was not particularly to train her to do the job but that using the MAPS prompter would be a cognitive orthotic and this task configuration would allow her to easily switch to new tasks with the only cost being the creation of a new script. This worked very well for her; the job coach said that the "training" time was cut in third. However, there were two problems the coach wanted to solve for the job trainee: (1) occasionally she would run out of work and it would be good for her to have another task that she could always do; and (2) when she decided that there weren't any more tasks to do the prompter offered a task that encouraged her "soft" skills (e.g., going to the supervisor and telling them you have run out of sorting to do). In the job coach's experience "Lacking this skill will "deep six our people" if they just stand there after completing a task. They need to be able to find something to do until they see their supervisor" and "90% of this is a soft-skill." In looking at what MAPS could provide to the situation, she felt that it would be difficult to put soft skills into a script. As a result, these last sub-scripts were quite critical; a frequent cause of people with cognitive disabilities loosing semi-sheltered jobs that are out in the public is a deficiency of these "soft skills." An example of this is the situation where the worker comes to the end of the task and just waits there, or arriving at work and not greeting the supervisor or other employees (Carmien, 2006b).

In this case the job trainee had enough cognitive ability to recognize the end of the task but not enough to initiate the soft skill activities. MAPS was modified (Figure 2.4) so that small initial prompts of subscripts were provided from which to choose. The ability to choose from a pictorial menu empowered the person with cognitive disabilities and motivated her to "own" the sub-task thus increasing the probability of interacting smoothly with co-workers and supervisors.

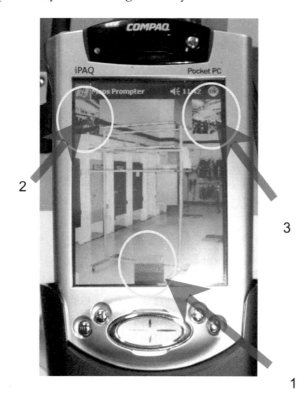

Figure 2.4: Selecting sub-scripts. In MAPS Prompter. From Carmien (2011).

2.1.3 CONCLUSION

This concept is key to designing AT for people with cognitive disabilities, because these tools can support the cognitive deficiencies the user brings in a way that exactly compensates just enough to get the task done. The act of making IA systems is more-or-less figuring out how to flexibly implement "just enough."

To a large extent, moving from an AI approach to an IA system is deciding where to place the divider between what the systems brings to the task and what the user brings to the task. By doing this optimally the person using the system can be engaged and feel genuinely autonomous.

Properly done, this increases the generalizability of the system in terms of different tasks, and leverages the user's abilities to recover from potential errors. The tricky part is to properly access what the user can offer, what must be supplied by the system at minimum to succeed and the fit between the two. These concepts will be discussed in Sections 4.1, 2.2, and, most fundamentally, 2.3.

2.2   DESIGN FOR FAILURE

Short Definition: Complex AT is by its very nature prone to functional failure or not achieving the intended goal. As a result it is necessary to design these systems with the ability to robustly accomplish their intended purposes by having the ability to dynamically mitigate such errors.

Longer Description: Design for failure (DfF) is a response to the inherent, but not obvious, flaws of complex AT that is used as cognitive support. The basic process is to detect an error, classify it using a "two-basket" approach for making decisions about whether to provide human intervention or to automatically mitigate the error condition, and finally intervene in the error with the appropriate mitigation strategy. This section describes a framework for classifying, capturing, and designing possible mitigation strategies. This framework is comprised of four types of failure events. Additionally, it provides a multi-step program for finding vulnerable parts of a system. This is a particularly long section, as it is not only intended to discuss the concept but also to provide an "algorithm" for creating a design for failure system.

2.2.1   INTRODUCTION

AT designed to augment cognitive ability has a history of fragility, difficulty of generalization (Carmien, 2011), and complexity (LoPresti et al., 2004) leading to a lack of wide adoption. Critical parts of the problem are the fragility of modern IT technology and the unbounded range of the application. Smartphones run out of batteries, network connectivity comes and goes, small devices can be easily lost or stolen, all leading to a complete halt of the assistive or cognitive orthotic support (Cole, 1997). Additionally, these systems are typically used in uncontrolled environments. For instance, there are traffic jams (Carmien et al., 2005) and unexpected complications of simple tasks. These situations, when looked at analytically, are usually solved with situated cognition (Suchman, 1987; Lave and Wenger, 1991), an ability that may be missing from the end-users of these systems.

In the CLever project (CLever, 2004), a series of interviews with parents of developmentally disabled young adults explored the use of hand-held portable devices to guide their children in using public transportation. Many caregivers expressed concerns about technology leveraging their children into situations that, if the support failed, would have catastrophic consequences (Carmien, 2007). While further exploring this topic (Sullivan, 2004), the CLever group collected many examples of people with limited cognitive ability "disappearing" in the public bus system, only to turn up many hours later due to becoming lost (Sullivan, 2003). Similar interviews with parents about multimedia script support of jobs for their cognitively disabled children led to their concerns about the system not being able to respond to unexpected situations in performing their jobs. In their experience, lack of immediate support often triggered a panicked response leading to employment problems (Carmien, 2007).

Why and how do these technologically complex and unbounded applications invite failure? The first and most common concern is that the technology used is not robust; it works in most cases but not reliably enough for the target population. Often, typically skilled end users are proficient enough to work around these problems or are able to triage error messages/failures, but frequently those users with cognitive deficiencies cannot overcome them. The systems themselves are built on complicated and layered fragile operating systems, services, and devices. It is common to experience computer operating systems halting, particularly in relation to networking. Further, the user population of the application is too small to guarantee thorough testing. Contributing to failure in mobile systems are connectivity issues that are context dependent. Examples of these are interactions between radio and buildings/topology leading to intermittent service. More frequent and with larger consequence are power issues when away from a charger, and running out of power.

The complex nature of these systems, having system server, mobile device, wireless connectivity, and unexpected environmental concerns (e.g., traffic jams, misidentified signs/tags, etc.) leads to having many failure modes. The support for the user and the user's interactions often need multiple sequences of correctly happening events/actions to succeed. An excellent example of this is the COACH (Cognitive Orthosis for Assisting with aCtivites in the Home) project (Mihailidis, 2007) which had to adapt to guiding people with dementia through many possible paths to correctly washing their hands. Furthermore, younger people often design interfaces for these systems. Without careful planning, designs are produced containing metaphors and affordances not obvious or right-sized/text/color contrast, etc.

Complex systems designed for the "real" world (in contrast to medical or sheltered workshop environments) are, by nature, unbounded. Real-world events are not nailed down, controllable, or predictable (with some statistical exceptions). The intersection of user, context, and device present a type of combinatorial explosion problem space of possible failure/error states.

Responding to this, I have developed an approach called *design for failure* (DfF). This is not a particularly new way of designing complex technology. The ubiquitous RAID systems (Patterson et al., 1988) that use a number of hard drives to provide reliable data storage is a design based on the assumption that hard drives fail. DfF is also an approach in software engineering in general (Sommerville, 2001). In this case, the approach means to spend a significant amount of design time on design focusing on the inevitable failure of the system. To obtain this, a fair amount of time must be spent on developing failure scenarios. This DfF approach can also ensure good design for the system when it does not go into a failure mode, not unlike the way well done accessible design can lead to better design in general.

This section will focus on some general principles:

- device failure;

- environmentally caused failure;

- failure due to user action; and

- failure due to supporting technology in special contexts.

Further examples from two projects (ASSISTANT (2012a) and MAPS (Carmien, 2004b)) illustrate the DfF approach.

2.2.2   DOMAIN BACKGROUND

The sorts of systems described here typically extend ability or attempt to aid in compensating for lost ability of people with cognitive disabilities. Therefore, this is a discussion of orthotics (Pollack et al., 2003) rather than AT per se. An illustration of the difference is a navigation system that supports your use of transportation and deciding which way to walk, but does not actually transport you there. Examples of these types can include navigation systems, task support systems (for work or recreation), communications systems, email and browser systems, and applications on PCs and tablets specialized for people with cognitive disabilities.

System failure means that an error has occurred in the expected behavior of the system, in this case an error in a socio-technical system. Socio-technical systems are composed of a user, a technical system, work practices, and the use context. Classical human error research (Reason, 1990) can give some insight into the problems with designing complex AT. However, the incidence of comorbidity (multiple disabilities) results in an effective "universe of one" as the design target (Erikson, 1958). The complexity of user baseline ability and daily and, in general, temporal variability of these abilities (Cole, 2013) make the assumptions that research based on typical or even highly skilled people difficult to extrapolate from.

Often the big difference between these technologies and more mainline AT is that obtaining the desired result only happens after a long string of connected events, and the failure at any one point (missing the bus stop to exit on) can cause the entire task to fail. The number of things that can go wrong (e.g., the grocery store has moved the contents of one isle to another side of the store) can make the error space very big.

Finally, the cognitive and experiential resources of the targeted user can be quite limited, making the system's use much more "brittle" than when used by a more typical population. An example of this is the use of a smartphone by an elder who may find accessing the on-screen keyboard tricky (too small keys on screen, popping back and forth between entry modes i.e., keyboard disappears without warning). This can become upsetting enough to trigger abandoning use of the application and perhaps any smartphone application entirely. This is magnified by the resultant AT abandonment (Phillips and Zhao, 1993) in exactly the population whose use of it may be life changing, not just convenient.

2.2.3   BACKGROUND

This study of DfF is illuminated from two perspectives: (1) human error studies and (2) the HCI version of distributed cognition (Hollan et al., 2001) and situated cognition/action (Suchman, 1987). Between them, these complex systems can be made more robust and adoptable.

2.2.4   ERROR

Modern, cognitive science-based studies of human errors start with Norman's (1983), Lewis and Norman's (1986), and Reason's (1990) work in the 1980s and 1990s. Their work on error classification (slips and mistakes) and turning an analytical eye to attention and memory as critical components supported design guidelines such as "forcing functions" and appropriate system responses to internally recognized error states. Dekker (2006) added to this the deep examination of error, describing it as systemic behavior. By seeing error as being comprised of the user, the system and the environment, designers can focus more clearly on capturing and mitigation of the error.

One of the more useful discoveries by these researchers is the actual distribution of errors in the world. The possible enumeration of incorrect actions and errors in any given task is a very big number, approaching the size of combinatorial explosion if you count all the elements and links that have to be present and happen correctly to perform even a simple task.

> *Fortunately, the reality is different. Human error is neither as abundant nor as varied as its vast potential might suggest. Not only are errors much rarer than correct actions, they also tend to take a surprisingly limited number of forms, surprising, that is, when set against their possible variety. Moreover, errors also appear in very similar guises across a wide range of mental activities. Thus it is possible to identify comparable error forms in action, speech, perception, recall, recognition, judgment, problem solving, decision-making, concept formation, interpretation and the like* (Reason, 1990).

Thus, it is more like a Pareto distribution, which frees the designer to work on the 20% of error types that constitute 80% of the actual errors.

2.2.5   INTELLIGENCE AUGMENTATION

The second important part of this problem is that these complex systems are designed to provide a cognitive orthotic for elders, children, and people with cognitive disabilities. These systems are designed not to replace intelligence or cognition but to leverage existing abilities to compensate for missing ones (CLever, 2004; Fischer et al., 2004). This is an AT variant of the approach started with Engelhard's description of intelligence augmentation (Engelbart, 1962). Intelligence augmentation (IA) in this domain depends on the correct application of distributed cognition. The capturing of the error state may or may not depend on the process of distributed cognition (i.e., the cognitive artifacts). However, mitigating the error within the system (in contrast to bringing in a helpful human) will involve generating and delivering such artifacts.

Again, the most appropriate (e.g., humane, require the smallest effort, and best fitted to the user) response is to support the user in their desires (Fischer, 2001b). This may be continuing to the end of the task or it may be safely and cleanly abandoning the task or returning home.

Strategy and Examples

The Two-Basket Approach

The discovery that the majority of the cases of error or failure are clustered in only several types of errors makes the problem tractable. However, there are still these events that can be identified as possible errors but cannot either be classified with confidence, may not be amenable to solutions, or may only be amenable to solutions that themselves are so complex that make them (the solutions) highly vulnerable to failure themselves. Those errors that can be classified and easily mitigated are put in one category and the (less probable) errors that cannot either be classified or, if correctly classified are not easily mitigated, into another. Fortunately, the second basket is statistically much smaller than the first basket (e.g., sleeping past the bus stop vs. a bus driver just leaving the bus). Therefore, for the second basket, the strategy is to bring in a human, ranging from a friend/caregiver to possibly emergency personnel.

Another simplifying strategy is not to focus on the cause but work with the error in a "stateless" fashion. Of course, if possible the end user and/or a caregiver should be informed that the phone battery is not holding a charge, but the important issue is the best way to help a user with a dead phone.

MAPS and ASSISTANT Projects

The two systems used as examples in this section are MAPS and ASSISTANT. MAPS, the author's dissertation project (Carmien, 2004a), is a system to support performing tasks by the user which cannot be done unaided. MAPS used a PDA-based prompter, guiding the user by displaying task segments that the user can perform. A set of these effectively leads the user through the accomplishment of the desired action. MAPS consisted of software to display scripts of prompts and collect performance data about use and a PC-based script editor that allows the creation/editing of tasks scripts using user supplied photos and recorded verbal prompt instructions (Figure 2.2).

The editor also provided utilities for loading the scripts onto the PDA-based prompter and collected performance logs. Supplementary functions of the MAPS system include a set of template scripts (>300) on a MAPS server on the internet and a prototype of an error trapping/help summoning sub-system based on queries embedded in each script prompt which, when triggered, sent SMS messages to a designated caregiver's cell phone. The system also includes undo functions, video-based help features, and an error-trapping visual programming environment.

ASSISTANT's (2012a) objective was to provide accessible support for the use of public transportation by older adults. ASSISTANT did this through an online application for trip planning and the provision of guides for multi-step journeys (Figure 2.5), with information provided to help the end-user to get from the last transit stop to the final destination. The main target group of the ASSISTANT project is mobile older people, aiding them in using public transportation by providing only relevant information, at the right time and in the appropriate format, to the user via audio, visual and haptic cues. ASSISTANT provided safety and security by providing error-trapping and remediation functionality. The user gets waypoint directions via prompts on a smartphone, both in written and audio formats, as well as with a map.

The ASSISTANT system identified the vehicle to board and signals to the user its arrival. The system also informs the user of when to exit the bus/metro using real-time location data. ASSISTANT provides a web browser-based route editor and personal preferences editor (Figure 2.1), a server that coordinates routes and checks for errors, and an application—a Personal Navigation Device (PND) (Figure 2.2)—running on a smartphone (Android or iPhone).

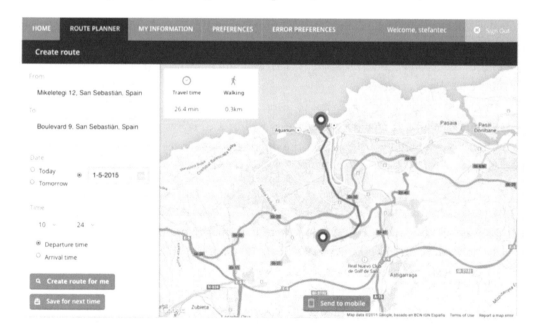

Figure 2.5: ASSISTANT route planner. From Assistant Project (2015).

Failure and Mitigation Cases

For each topic there is a section describing the error's cause, how the error is recognized, and the mediation the system applies. Note that the cause section does not describe the chain of causality that produces errors, but describes the proximate cause of the problem.[6]

Device Failure

Cause: It is increasingly common to find support systems of this type being composed of a server of some sort and a mobile device. Server failure is a fairly uncommon (shy of original software bugs) occurrence and can be reduced further by hardening the device (i.e., uninterruptable power supplies, RAID redundancy, etc.) or moving the service into the cloud. However, both the mobile device and the connections to the mobile device can be quite fragile. Accordingly, there could be a dead battery, a lost phone, a stolen phone, and any number of things that might make the phone inoperable by the end user. The connection between the phones may fail too, there may be no phone service, there may be no GPS service, or worse yet there may be no data connectivity between the server and the phone. Finally, the phone may be partially broken—the screen may break, the Bluetooth earpiece may fail, etc.—hardware problems of all kinds that allow some phone functionality (typically as just a phone) but not other critical abilities.

Capture: The capturing of error states in the device failure category typically requires repeated polling of the possible failure points. In the ASSISTANT project, this polling happens on both sides of the vital mobile server connection. Therefore, at the smartphone side the ASSISTANT application monitors the status of the GPS service, the phone, and data connections and the battery level (Figure 2.6). On the server side, there is a periodic polling (ping) of the smartphone's connected state and part of every data exchange is the status of the battery charge.

---

[6]   This is actually a very relevant point. One of the common mistakes that beginning developers of AT make is to start with diagnosis as the ordering principle in generating requirements and eventual design. This is not as effective or generalizable as the more common functional approach, i.e., designing to compensate for a missing facility rather than for people with a disease that may cause that missing ability.

Figure 2.6: PND notification screen. From Assistant Project (2015).

By collecting this data the system can flag connectivity and hardware problems in real time (Figure 2.7). However, it is important to discriminate between intermittent problems and real problems. This is particularly important to prevent type 1 errors, i.e., detecting and responding to an error when there was not that problem. False positives can contribute to AT abandonment (Kintsch and dePaula, 2002) by frustrating the user with false alarms.

Mediation: In the case of ASSISTANT device/connection failure there are, like the sensing, two sides to the mitigation activities.

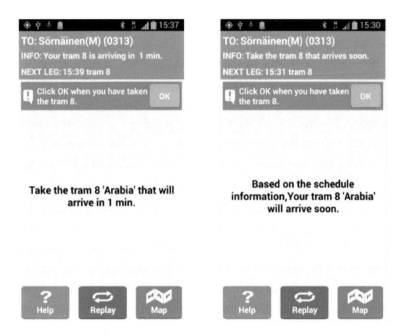

Figure 2.7: Navigation guidance (left: with GPS, right: with no GPS). From Assistant Project (2015).

On the server side, when the pinging of the smartphone fails over a predetermined period of time (to accommodate temporary intermittent disconnections), the system sends a SMS message to a contact the end user has specified. This is also called a "dead man switch" in other engineering contexts. The un-implemented part of this design in ASSISTANT specifies that the message be sent to the first person in an escalating list, if that first person, say a cousin, does not send a SMS reply to the sender in a specified time the system then sends the same SMS to the next contact, and so on until a contact replies with a SMS that indicates that the message has been received and that they will be responsible for the user with the disconnected phone. The last entry in the contact list is usually an emergency service such as the police of a private contractor. In this way there is a safety net below the user that will always provide support should things go terribly wrong. Note that the end user can choose not to have this function at all or not to call in the police or an ambulance service; this can be specified in the preferences pane of the route editor. The content of the message has in it as many specifics as the system knew at the time of disconnect so that the called person can then provide appropriate help. An example of the SMS is: "*Hello, this is <my name>, it is <time> on <date>. I am sending this to you at <time> on <date> From my phone <phnumber>. My location right now is <currloc> and I am going to <dest> on <last bus taken> and I think I am in trouble. I need some help.*" If the server determines that this is a case of the battery running out, this is included in the message. This single contact SMS notification is currently implemented and tested in the ASSISTANT system.

The PND informs the user of low battery conditions with a dialogue box, and every screen in the navigation set can display the loss of data, GPS, or phone service to the user. When the real-time vehicle location data provided by the public transportation administration is not available, the navigation screen informs the user that it is now giving directions based on the route schedule and not real-time data (see Figure 2.7). By giving these updates of potential problems the user can figure out a workaround to the loss. Alternatively, by pressing a button, the user can have a SMS message sent to contacts from the phone with appropriate status information. The end-user also has the option of calling the contacts via the phone directly with a push button on the help screen.

Environmentally Caused Failure

Cause: Another mode of failure, outside of the user's actions is when the environment changes unpredictably, in a way not expected either by the user or tacitly by the task execution sequence.

An example in the ASSISTANT system may be that the route has been changed at the last minute by the transportation agency. This may not yet be reflected in the transportation systems database. Perhaps there is an accident or some unexpected construction or a very popular event has caused a traffic jam so large that the driver has decided to drive around the entrance to the stadium rather than in front of it.

For a user of the MAPS system the problem may occur when the arrangement of the cleaning supplies for the job that the MAPS prompter guides the person with cognitive disabilities to do (which was too complicated to do without) has been moved.

Capture: In ASSISTANT the PND and its server monitor the position of the user's smartphone and the location of the vehicle. It continuously compares that with the expected location and time according to the route that was loaded into the PND from the server at route creation time. Additional deductions from this can be made from this data. For instance, if the location information indicates that the PND is moving very slowly (0–3 MPH) when the user is expected to be on a tram (10–25 MPH) the system can deduce (with due concern for the problems type 1 errors can cause) that the user is now walking or standing still.

In MAPS, the log of the user stepping through the series of multimedia prompts that guide her to do a task may also have time stamps. MAPS can conclude that there is a problem if the user stays on one prompt for a much longer time than would be broadly normal for this person to do for this segment of the task, or (more interestingly) the user is "rocking" back and forth over two or more prompts, centered on the step that is to be accomplished at this stage of the task. The experimenters also observed the user going through the task steps very rapidly, much more rapidly than could be accounted for by actually doing them. It turned out that she had memorized these steps (from previously cooking exactly these cookies in her special education classes many times before) and one of the cooking instructions triggered her performing the next five steps. As a result, she just smoothly did them all at once. While not an error, in that the guided task was accomplished, it did signal that this particular 36-step cookie cooking script could be, for this particular user, condensed to incorporate the "skipped" steps into the one cumulative step.

Mediation: ASSISTANT, when it determines, by real-time location and velocity tracking, that an error state has occurred, it starts the error recovery process. This may mean (1) instructing the user to exit at the next stop and recalculating the new best route to the goal, (2) informing the user of what is going on, (3) pushing the new route to the PND, and (4)starting the route guidance. In designing this mitigation, the design team first struggled with elaborate plans for both guiding the user back to the place where their route differed from the planned route and then getting them back on the original track. This was both pragmatically difficult and hard to represent to the user. However, after several attempts it was discovered that it was much easier (and able to reuse code and GUI to do so) to simply generate a new plan. The new route was calculated starting from where they were now, and was aimed at the original target destination. This was implemented and tested in the field prototype with positive results. If the user set in the preferences panel of the route planner that they are uncomfortable with this happening or with any new route that has more than n steps (i.e., takes three different buses and a metro ride to get to the goal from where they are now) (Figure 2.8), the server initiates a SMS message to a trusted contact to resolve the problem. This resolution may involve "rescuing" them or just getting in touch with a contact by phone for some reassurance.

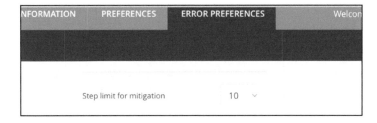

Figure 2.8: Setting the users preference of the maximum number of steps in a route mitigation. From Assistant Project (2015).

Failure Due to User Action

Cause: This category contains a larger percentage of difficult to capture and categorize errors than the others. Users can be very clever when trying to accomplish tasks. These tasks may be a job or navigating routes. In doing so, the mismatch between metal models and environment/device may generate actions that are legitimate attempts to move forward but errors nonetheless. While these errors, when captured, may frequently fall into the second categoty (those whose remediation requires human intervention) the research-based statistics of clustered mistakes still holds, and the system can still be quite effective. Sometimes the user is unable to make a decision about which way to proceed in following the route instructions. In interviews with Colorado Front Range travel trainers for young adults with developmental disabilities (Sullivan, 2003), the most frequent scenario brought up was sleeping through their bus stop. Another common concern is that the environment would be so noisy that the passenger might not hear their stop being called out. In the MAPS project a user could take too long in doing a task (Carmien and Gorman, 2003) or in a job requiring moving through a building (e.g., delivering corporate mail) the MAPS prompter may not be triggered by proximity to a Bluetooth beacon at one of the mail delivery stations (Gorman, 2005).

Capture: The ASSISTANT server and the PND both monitor and compare location and velocity information with what should be happening if the correct route was being followed, especially when correlated with vehicle real-time location information. Capturing the error state this way gives not only the existence and type of the error but also the current location and time of the user.

In the MAPS system, a prototype had a time that each task in teh script was expected to take. When the advancement to the next step prompt did not happen for a significant about of time over the expected task duration an error was flagged. A prototype of the caregiver editor provided a set of screens to build up an error trap and corresponding mediation plan, as shown in Figure 2.9.

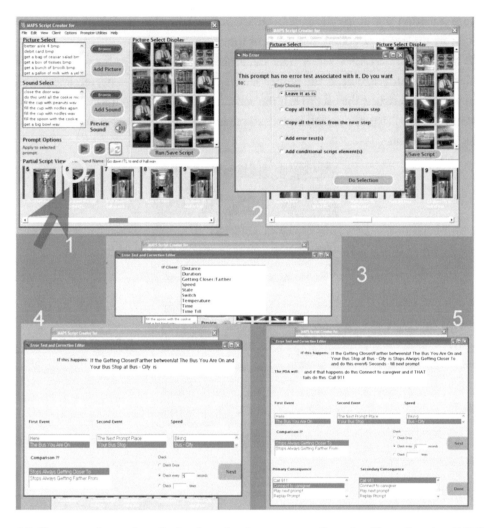

Figure 2.9: Error trapping and mediation insertion into prompt of a script. From Carmien (2005).

Mediation: For some user errors a good way of mediation is for the system to provide better ways to support the user in making decisions. So ASSISTANT, which is a waypoint-to-waypoint navigation guidance system, also provides a map to both inform the user of their current location and also reassure them about their progress on the route. Often the problem is not with representation of the route to the user, but a result of a lack of attention to the PNDs instructions. To mitigate the lack of attention the PND can be programmed to vibrate when the next step on a task is about to be presented. The use can also choose to have the screen blink in a similar situation. The default mode of prompting for the route instructions is through a Bluetooth earpiece, which does not require continuous attention that instructions on a smartphone screen do—a push of information rather than requiring the end user polling the status of the screen.

When the ASSISTANT server does determine that the user is, for whatever reason, not continuing on the correct route at the proper velocity, it flags an error state, recalculates the route, and, if appropriate, instructs the user to exit the vehicle. The user can also initiate this process via the help screens (Figure 2.10) of the PND application, and on the same screen, the user can summon a contact to help them.

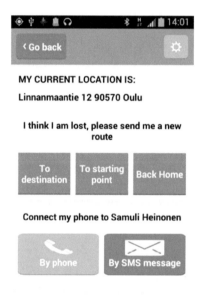

Figure 2.10: PND help screen. From Assistant Project (2015).

MAPS sent SMS messages with the location of a step in the task that had been not reached within a caregiver/end user specified amount of time. Similarly, skipping steps or taking too long to accomplish a task can (depending on the user's preferences) trigger redisplay of the missed task step.

Failure Due to Supporting Technology in Special Contexts

Cause: High complexity navigation or task-guiding applications are embedded in a matrix of supporting and helping applications and systems. For instance, ASSISTANT relies on WebServices for messaging and remote procedural calls between its modules, on GPS for location and velocity information, on SMS for contacting help 4G mobile data service for connecting to the ASSISTANT server, and on Bluetooth to connect with an earphone. All of these have different levels of reliability, range, and sensitivity to environmental conditions.

One of the "killer applications" in the 21st century is the various GPS systems offered for automotive and other vehicles. These, combined with digital maps and simple AI-based optimal route generators, have changed the way that we drive and travel.

They are reliable (depending on the correctness of the map used) and widely used from family vacations to taxis to police vehicles (and of, course by the original users, the armed forces). However, commercial units are not reliable for pedestrian urban navigation (Weimann et al., 2007; Andersson, 2012) . This is mostly due to the well-known canyon effect. In the process of designing the support system for the "last KM" (the trip from the last transportation stop to the door of the route's goal), the design team in Helsinki ran tests (one example is Figure 2.11) that led them to design around the use of GPS in the standard way.

Additionally, in ASSISTANT, while the transit systems did have reliable GPS systems that were coordinating with their administration centers, and the transit systems provided ways for us to utilize that information, sometimes the real-time location of the transportation systems vehicles are not available.

Figure 2.11: Actual urban route taken vs. smartphone GPS. From VTT (2013).

Capture: This is a good example of the two typical approaches to working with this class. In some cases the best approach is to work around the problem. That is to say, by avoiding the standard use of the untrustable technology. In other cases, capturing and providing mitigation is more appropriate. An example of the first approach was avoiding altogether the typical urban use of GPS and an example of the second approach was having the design sense the lack of real-time data when the calls to the real-time data server returned an error message, and then providing a workaround.

Mediation: ASSISTANT designers acknowledged the unreliability of existing GPS-based navigational aids for pedestrian navigation guidance, but knew that there was useful knowledge that the GPS gave and were really motivated to produce something for the elders that was reliable and would avoid giving any guidance that might frustrate the user and trigger abandonment (Kintsch and dePaula, 2002). The solution was three-fold.

1. Provide an appropriately sized map for the user from the last stop to the goal. This was possible because the system knows where the user is when the bus or metro has reached their last stop and they have acknowledged debarking (Figure 2.7, left).

2. Provide a compass pointing to the last goal during the walk from the stop to the door of the route's end. How this was done was averaging the GPS signal over time and knowing absolutely where the end goal was and having a reliable compass in the smartphone. The drawback is that the compass pointed not at the next waypoint but directly at the front door of the goal, but this was carefully explained to the end-user in the accompanied short user's manual (Figure 2.12, right).

3. Use this same averaging over time technique with the GPS position to determine if the user is walking the wrong way and alert them of this with the smartphone. Again, it was important to gather average locations and choose a span of time (typically one to several minutes) that they were going the wrong way so that small movements were not misinterpreted as being confused about where the end goal was. This was particularly important as working with the compass may require backtracking to walk around buildings or other obstacles, again driven by the line between providing useful timely guidance and raising false alarms that contribute to system abandonment.

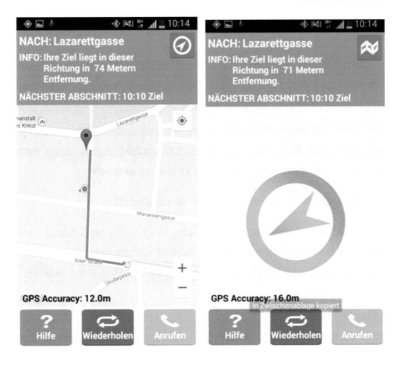

Figure 2.12: Last KM guidance (left: map view, right: compass view). From Assistant Project (2015).

The other problem in ASSISTANT was addressed by changing the guidance screen with adding text on the screen alerting the user that the schedule is being used, and the user manual advises the user that when that happens they need to be a bit more flexible in terms of planning the next steps (i.e., the bus may arrive in something like 3 min). Field tests so far show this to be small enough not to compromise the reliability of the system.

Developing a Culture of Design for Failure

A culture that supports DfF sees failure as a whole system attribute not only a human problem. Dekker points out that we need to move from seeing error as the cause of a mishap to seeing that mishaps and failures are symptoms of deeper, systemic, trouble. He calls this the *New View of human error* (Dekker, 2006) and one of the marks of this approach is seeing complex systems as not intrinsically safe, in contrast to the more traditional view that complex systems are basically safe with human error undermining that basic safety. With this figure ground reversal it becomes much more natural to find the points of potential failure.

Design for Failure with Scenarios

DfF works well with a scenario approach (Rosson and Carroll, 2001), but the traditional scenario design approach needs to be extended to include branches in action. It needs to anticipate, at each critical point, where things often go wrong, how to capture, and how to mitigate. Here, domain experts become invaluable, both in identifying critical points and estimating the relative probability and consequences of failure.

DfF error scenarios can be generated by enumerating failure points from anecdotal narratives, from end-users and proxies, and from literature/best practices collections (among other sources). The basic outline of our approach has been:

1. prioritize failure importance from probable frequency, consequence (i.e., danger) probabilities, context (i.e., time and place), and results (i.e., confusion matrix and type 1 errors lead to abandonment, type 2 errors lead to danger);

2. determine ease of detection for each error scenario; and

3. based on 1 and 2 sort into the two baskets. In general, high consequence, low frequency and/or hard to detect events are sent to human support and high frequency, easy to detect, easy to mitigate, and lower consequence are good candidates for automatic error correction.

A DfF approach adds a requirement for a module to monitor and continuously evaluate the progress of a system-supported activity. Additionally, the system should be easy for the end-user to autonomously generate an error flag, as well as provide any summoned help context information (and possibly a log of the user action, depending on privacy issues).

Table 2.1: Mitigation by category

Category	Mitigation Approach
Device failure	Dead man switches—constantly monitor the critical device (e.g., ping every 20 s) and if no response in reasonable time automatically summon help
Environmentally caused failure	Notify user when captured, replan guidance if possible, or else notify contact person by phone or SMS
Failure due to user action	Offer to rework guidance (e.g., "re-calculating route" on automobile GPS)
Failure due to supporting technology in special contexts	Use what parts of the technology that are functioning and provide as much decision support as possible (i.e., "there is a printer within 3 m of you, it may be inside the room you are outside of")

One nice property of incorporating DfF into the system from the beginning it that it lends itself very nicely to iterative improvement. Once status monitoring, error trapping and mitigation modules are implemented it is easy to include increasingly sophisticated error capture and mitigation technology, starting with being rule based and progressing to incorporating real-time machine learning approaches.

DfF adds a level of a robustness to complex cognitive AT as well as removes a common trigger for abandonment (which is the fate of up to 60% of high functioning AT (Reimer-Reiss, 2000)). But the few examples here show just how difficult implementing this approach is, not just in design but also in system architecture. Moving from well-bounded environments (interacting with just the screen and the network, without relying on external context) to dynamic AT in the wild will require both careful examination of what can be done to support existing skills of the user, what can be done by the system, and what will require external human assistance (Table 2.1). Further research will increase the percentage of the first two, but there will always be a need for the last.

2.2.6   CANONICAL PAPER

As I have pointed out, the idea of DfF is not new to engineering in general, however as a bounded HCI problem, there are few publications that focus on this principal. A good introduction to this is in classical engineering terms is Patterson et al. (1988).

2.2.7   AT EXAMPLES

The discussion of the DfF approach in the ASSISTANT and MAPS system above are two examples of how design for failure works and how to design with DfF.

The COACH Automated Prompting System

The Assisting with aCtivites in the Home (COACH) system (Mihailidis et al., 2008; Hoey et al., 2010) (Figure 2.13) is a Cognitive Orthotic system, e.g., a system that leverages existing skills to preform tasks that the user cannot do unaided or unguided. Within the spirit of IA most of the systems described in this book are orthotic (aiding the user to use what they have) rather than prosthetic (replacing what the user does not have). For example, contrast back or forearm-wrist braces with artificial legs or hands. This prototype is a result of a multi-discipline team and a reasonably successful third generation that took them from 2002 to 2010 to develop. The COACH system is an excellent example of how many different domains and person hours it takes to actually make an IA cognitive orthotic that works. The most effective (and needed) supports for independent aging and efficient use of caregivers are those categorized as activities of daily living (ADL) (and also Instrumental activities of daily living which support ADLs). COACH is designed to support proper handwashing by people with dementia in a home environment.

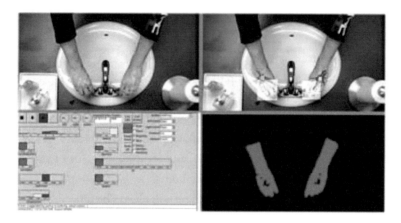

Figure 2.13: COACH system handwashing. From Mihailidis (2007).

The COACH system is composed of three modules: hand tracking, planning, and prompting (Figure 2.14). The hand tracking module uses video images (e.g., an overhead camera mounted above the sink in a washroom) to identify the position and actions of the user's hands and objects. Planning module translates the image information into washing progress information using a partially observable Markov decision process (Hoey et al., 2010). Finally, the prompting module, if directed by the planning module, provides guidance, in appropriate visual and audio formats, to the hand washer. The selected level of assistance is determined by an estimate of current user's lucidity and responsiveness provided by the planning module. COACH in use is continuously guiding the end-user, and the planning module can accept several paths or sets of sub paths for correct washing, there are multiple correct paths in COACH washing (Mihailidis, 2007; Czarnuch and Mihailidis, 2012; Czarnuch et al., 2013) (see Figure 2.15)

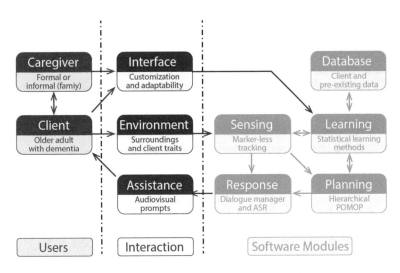

Figure 2.14: COACH modules and functions. From Mihailidis (2007).

This ability to detect the path and respond to anyone of several correct actions, missing one part that can be done in a different order, makes COACH an excellent example of designing for failure from the beginning of the project. The actual intention was to have people that really needed it and were not particularly using this to train in proper handwashing but to support this IADL behavior while the dementia progressed, up to a not reliable point.

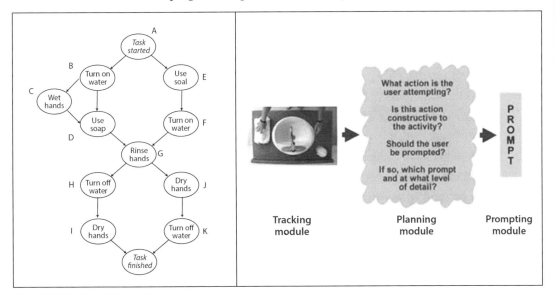

Figure 2.15: COACH plan and steps (left) and COACH modules (right). From Mihailidis (2007).

2.2.8 CONCLUSION

Design for failure is not really much more than an extension of requirements gathering combined with appropriate system architecture modularity. The main message is that systems, when used by people without a strong ability to compensate for troublesome situations, should anticipate common problems and be ready with mitigating actions. These can range from automatically reconfiguring interfaces, responses, and plans for the user to bringing in human assistance of appropriate urgency. While the need is most obvious in the examples described above, some awareness of this approach should be brought to any computationally supported system.

2.3 DISTRIBUTED COGNITION

Short Definition: Distributed cognition (DC) is the view that cognitive acts are (typically) based on both the internal assets of the person and the cultural structures and artifacts that support intelligence or cognition in a given human action.

Longer Description: Gregory Bateson remarked that memory is half in the head and half in the world (Bateson, 1972; Pea, 1993). We exist in a world full of examples of this distributed cognition: the shopping list that "remembers" for us, the speedometer on our car, the position of the toggle on our light switch (up for on), the very words that we are reading right now. Acts and knowledge are not constructed unilaterally. An interesting question is "Where is the boundary between my knowledge and the context that supports my knowledge?" (Salomon, 1993). Distributed cognition is an approach that views the cognitive act as a result of a system comprising an actor (memory, attention, executive function), the artifacts in the actor's environment, the context, and possibly other people. These artifacts can be as concrete as a notebook and as ethereal as language. Viewing cognition in this fashion can enable analysis and prediction of cognitive behavior that has a basis beyond the solitary human mind.

Distributed cognition attempts to analyze problem-solving behavior with a unit of analysis that spans individuals, artifacts, and others (Hollan et al., 2001). The artifact can provide external support (i.e., amplification and transformation) for cognitive acts that may be beyond the ability of the unaided mind (e.g., cube roots). The artifact may be providing a true cognitive orthotic role, as in the MAPS prompting system, or may just extend sensory abilities as in the classic blind man's stick in Gregory Bateson's example:

> But what about "me"? Suppose I am a blind man, and I use a stick. I go tap, tap, tap. Where do I start? Is my mental system bounded at the handle of the stick? Is it bounded by my skin? Does it start halfway up the stick? Does it start at the tip of the stick? (Bateson, 1972).

Distributed cognition is a cognitive science model, in the sense that it basically is concerned with the individual's internal cognitive processes and the support/extension that artifacts can provide, in contrast with the sociological/ethnologist view that sees the user and artifact as part of a system of relationships (Suchman, 1987). One view of distributed cognition is that it is attempting to describe how distributed units are coordinated, how information is represented, stored, and transformed, and in turn how the representation of information transforms the task at hand (Pea, 1993). In this sense, the representation and computational mechanism that manipulates the representation become part of the cognitive process. But this transformation is not a static event.

> *In distributed cognition, one expects to find a system that can dynamically configure itself to bring subsystems into coordination to accomplish various functions* (Hollan et al., 2001).

Therefore, the system is often dynamically interacting between the agents and objects, each modifying and mutually supporting the effort toward the system goal.

From the distributed cognition perspective, this process of "external cognition" (Carroll, 2003) consists of agents creating and using information in the world, rather than simply within their heads, to do three key things: (1) externalize information to reduce memory load (such as reminders); (2) simplify cognitive effort by "computational offloading" onto an external media; and (3) allow us to trace changes, for example over time and space, through annotation (Perry, 2003). The external cognitive artifacts or mediating artifacts that support this offloading increase memory capacity; in addition, the representation held in the artifact may "not simply augment, or amplify existing human capabilities. Rather, they transform the task into a different one" (Norman, 1993).

To analyze a task or environment from a distributed cognition perspective one needs to answer three questions (from Pea, 1993):

1. *What is distributed (i.e., different components of the problem-solving process as well as the product)?*

2. *What constraints govern the dynamics of such distributions in different time scales (e.g., microgenesis, ontogenesis, cultural history, and phylogenesis)?*

3. *Through what reconfigurations of distributed cognition might the performance of an activity system improve over time?*

The process of deconstructing the problem with this framework can be useful in creating a system that distributes knowledge in the world (Norman, 1990) by redistributing expert skills into a system. In this case, what is distributed are mnemonic and executive triggers and content, the constraints on the system are the timeliness and fit of the prompts to the current context, and the improvement of the performance over time maps to both error correction and scaffolding concerns.

By viewing the cognitive system as a system comprising an actor and mediating artifacts with the perspective of distributed cognition, one can look at goals and plans to attain these goals as being effected by a system comprising actors, singly or in groups (e.g., classes of actors), mediating artifacts, and their interactions. There is no particular bias in this perspective toward human actors; all elements are evaluated on the same plane (Nardi, 1996b). Distributed cognition looks for cognitive processes wherever they may occur, on the basis of the functional relationships of elements that participate together in the process. In distributed cognition, one expects to find a system that can dynamically configure itself to bring subsystems into coordination to accomplish various functions (Hollan et al., 2001). Distributed cognitions perspective is biased toward a goal, the cognition; in contrast to focusing on the process, as in activity theory.

2.3.1   EXAMPLES OF DC IN OUR DAILY LIFE AND MARKS OF DC

The gaining of the view of distributed cognition (DC) is like owning a Saab—once you get one you see examples everywhere. For some this is a brilliant insight, for others it seems like just common sense. In this book I try and explicate these ideas as not only explanatory or theoretical principles, but to show that using them gives genuine guidance in the design process. In this case, by understanding and examining distributed cognition you will be led to look closer at the artifacts and the division of effort that DC implies. So as I was exposed to these ideas in L3D I would ask myself, "what difference will this make in my work?", not "Wow, another fascinating insight" or 'This is useful for classifying these things (systems, events, actions, etc.)". If it does not make a difference in what you are doing, it probably does not belong here.

The classic example of DC is the transition from memorization of epic poems (i.e., The *Iliad) and religious texts* (i.e., the *Quran*, the *Vedas*) to writing and reading; here the *this side* of the DC is my ability to read and the *that side* are books. Maps are the same; before you had memorized trails you did not need to know how to read a map, however, now knowing how to read a map you can use that map to "memorize" any trail. Similarly street signs allow us to locate ourselves in new places without memorizing the organization of a city or village. Of course the premier example of DC is the computer.

There are several important attributes of DC that the designer needs to know (Table 2.2). The first is that distributed cognition changes the task. This is obvious from our book example, that before reading the task was using all sort of mnemonic tricks to memorize a few large texts, but with writing the task becomes how to decode the letters to make words (in my head[7]). Armed with the ability to read, a person can now walk into a library and effectively have memorized (because of her ability to read and the collection of books) thousands of texts. So, the gaining of the knowl-

---

[7]   Interestingly, enough reading silently was not a skill that many had until the early Middle Ages, at least in the west. Augustine's confessions mentions that he was surprised when he came upon Saint Ambrose reading silently one afternoon in 384 AD, Manguel (1996).

edge to do this sort of cognition also changes, to teaching reading. Another example of DC is the transition of grocery clerks from reading the prices on the grocery items to scanning them with barcodes. This changed the task from picking up the can (for instance) and turning it to see the price sticker and then pushing the keys on the register in that amount to passing the can over the scanner and listening for the feedback beep of success. As with many transitions to DC there are also unexpected side effects that can be quite beneficial—now the consumer (and the store) have a record of what was purchased because the database entry corresponding to the bar code also has descriptive data that can be used.

Along with changing the task, one side effect is deskilling the user of the previous ability that is being replaced. A personal example is our family's use of GPS on family vacations. Previously we poured over maps to plan the route to be driven, and because of this had a pretty good map of the route, and often a rudimentary layout of the cities on the route, before we left. Now with accurate and reliable automobile-mounted GPS systems, it is not unusual to input the goal, have the system calculate the route, and leave only knowing the distance and time of arrival. On one trip from Bonn to a hotel outside of Paris the GPS indicated that we should take the next off-ramp and then we would be a half .5 km from the hotel. Unfortunately, the exit was under re-construction and as we passed it the GPS just recalculated the route and we found ourselves 20 min later at the same exit. We did not have a map of the area so, being the computer scientist that I am, I drove 30 min at right angles to the way we were going and then followed the directions; we arrived at the hotel from the other direction. Had I had a map I would have seen that the following exit would have allowed us to come back on side roads.

Another example of deskilling is the controversy about the use of calculators in primary schools. There was much concern that allowing them would deskill the children's already gained arithmetical ability. In 2011, The National Council of Teachers of Mathematics decided to recommend allowing their use in the classroom,[8] however (as with the slide rule), it was now very important for students to learn estimation skills to ensure that the calculator's seeming accuracy was at least in the right "ballpark" of the right answer. This may be an example of an auxiliary skill that could be required to ensure that the DC system does not make things worse.

For many implementations, particularly in the domain of supports for people with cognitive disabilities, it is important to fit the external support to the particular, perhaps unique, existing abilities of the end-user. And those abilities must be sufficient to effectively utilize the support system. Figure 2.16 shows how the fit can be critical for adoption and use. In the upper region are those people who could use the system but their cognitive skills are high enough so that using the system would be more work than not and produce the same result—they don't need it; in the lower region are those people that really need the system but don't have the resources sufficient to use the DC artifact, so they need it but can't use it. Those in the middle section can use it, and do need it. The

---

[8]   http://www.nctm.org/Standards-and-Positions/Position-Statements/Calculator-Use-in-Elementary-Grades/

tricky part is to make sure that the middle section is not too narrow so that the designer ends up basically developing bespoke software. The workaround here is personalization. Note also that the population in the middle may change over time, for instance people suffering from Alzheimer's may decline in ability over time so rapidly that by the time they have mastered the DC artifact they can no longer benefit from the system.[9]

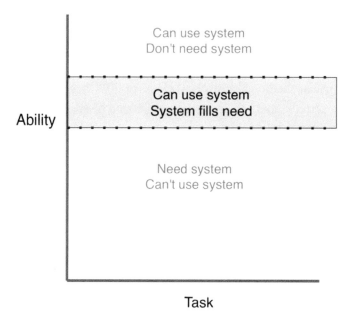

Figure 2.16: Bandwidth of need and ability.

[9]   Fortunately the typical course of Alzheimer's tends to shelve out for several years (Reisberg et al., 1996) when both the skills and the need for a system like MAPS to have some use.

Table 2.2: Marks of distributed cognition

Factor	Details	Examples
Reliability	All parts of the DC system must be highly reliable	• GPS in urban areas suffer canyon effect • Intermittent internet/mobile data
Cost of new skills < cost of old way (over time)	The requirements of learning the new DC required skills must be low enough to justify adoption	Square root on calculator << than doing it with and algorithm and arithmetic
Cost per use must be << than old way	Many tools for doing tasks that are easily mastered are never adopted	Doing simple addition or multiplication is typically easier than finding and using a calculator
Cost for failure of DC must be reasonable	If the consequences for failure are disastrous DC may be a bad choice	Early automatic driving cars
The inevitable deskilling should not leak into other activities		

2.3.2   CANONICAL PAPER

Distributed cognition explicates that cognitive phenomena generally are best seen as distributed processes. The theory does not do away with the notion of individual cognition but emphasizes studying instances of socioculturally distributed cognition: how cognition is distributed across people and artifacts, and on how it depends on both internal and external representations. Hollan et al.'s paper "Distributed cognition: Toward a new foundation for human-computer interaction research," while not the first discussion of this approach, is the one most pointed to (Hollan et al., 2001).

2.3.3   AT EXAMPLES

Memorize task steps vs. the MAPS prompting system.

2.3.4   EXTERNAL AND INTERNAL SCRIPTS

Scripting can be seen as an instance of distributed cognition. Cognitive scientists look at knowledge representation, particularly operational knowledge, in terms of scripts and frames (Schank and Abelson, 1977). In the MAPS system's view of scripts, however, they are regarded as exterior and supportive rather than as internal structures. Traditional rehabilitative use of scripts is intended to lead to the memorization of the script steps, thus tying together these two perspectives. In the MAPS system, scripts are designed to be external supports when the internalization of the sequence of instructions is not possible. From this, one can discuss internal scripts as sequences of behavior that have been memorized and can be appropriately evoked to accomplish a desired task, and external scripts as the distributed cognition artifacts to simulate an internal script (Carmien et al., 2007). Figure 2.17 illustrates the external cueing of extant "atomic" behaviors by an artifact or human support. The top portion refers to a person with sufficient internal scripts to accomplish the whole task, the bottom two sections of the illustration refer to a person with all the "atomic" behaviors to accomplish the task but not having the internal scripts to tie them together in sequence and detail. The middle section demonstrates a case in which the external support is more than the person needs to accomplish the task, therefore possibly creating confusion or boredom (see Section 2.4). The bottom section shows the right level and fit of external support.

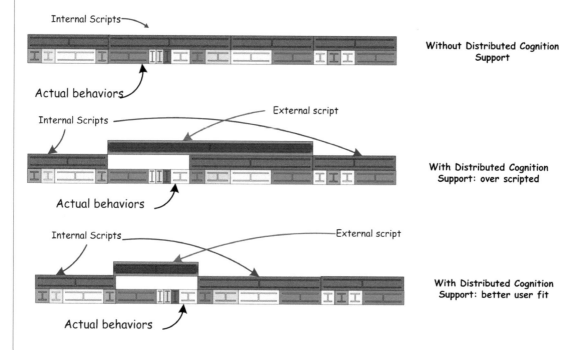

Figure 2.17: Internal and external scripts. From Carmien (2006b).

As an example of the process of internal scripts, consider when children become old enough to dress themselves that the various executive and mnemonic tasks involved with selecting, donning, and fastening clothing become part of an internal script that can be appropriately "run" when required. For some people with cognitive disabilities, the various internal scripts involved in the task of going to the store to buy milk may not be available; perhaps all the components but the travel component exist and are appropriately accessible. MAPS may provide an external script, in the form of prompts, to use the bus to the store, to accomplish this whole task.

Figure 2.17 demonstrates the relation of these internal and external scripts. Even people suffering from severe cognitive disability have functioning internal scripts for simple functions such as eating or walking. The MAPS system envisions its external scripts as bridging the gaps where the internal scripts do not support the complete task behavior.

2.3.5   CONCLUSION

DC is the understructure of all AT/IA systems. Shifting the line that divides what is brought to a task by the end-user, who may not have the resources required, and providing computational support for the missing abilities, can level the playing field. Even if the system does not explicitly look like an example of DC, it may provide very useful insights into the problem of spending a bit of time identifying what parts of the problem are based on this concept.

2.4   SCAFFOLDING

Short Definition: Scaffolding is the support given during the learning process which is tailored to the needs of the student with the intention of helping the student achieve his/her learning goals (Sawyer, 2006). Scaffolding can also be part of task support and dynamic scaffolding can be a personalization fit for AT systems.

Longer Description: From educational domain, notion is that the scaffolding provided support for the difficult parts, allowing the existing skills of the student to achieve success early in the process, and as the learning process continues the scaffolding can be slowly eliminated until the student has mastered the skill being studied and can accomplish them without help. An elaboration on this concept is extending or dynamic scaffolding, a recent example of which is smart cars automatically pulsing brakes when skidding.

Supporting human goals with computational artifacts must take into account the variability of human need and have a goal of fitting them, one measure of which is the concept of flow (Csikszentmihalyi, 1990) (Figure 2.18). The notion of optimal flow is that humans perform most comfortably in the zone where the task is not too difficult or too easy. When a task it beyond current ability there is a danger of abandonment; when it's too easy then it may become boring (and also abandoned). The original work by Mihaly Csikszentmihalyi was with typically abled adults, but it can be extended to our population, and problems ameliorated by providing scaffolding stuctures to compensate for either condition. Dynamic on-the-fly or static history-based scaffolding can reflects flow optimization. As skills change then challenges and tasks change.

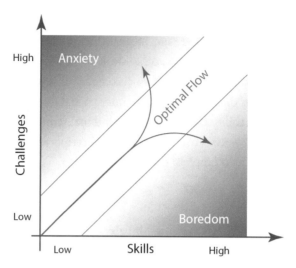

Figure 2.18: **Flow.** Based on Csikszentmihalyi (1990).

Therefore, one way of providing an optimized set of supporting tools is to provide scaffolding that fits the user's model. Doing this employs changing the interface and supporting functionalities as the user model changes. This model reflects the user's current skill set and needs, examples of which are the work done teaching programming, learning the scientific method, and guiding collaborative learning. A somewhat failed example of scaffolding was the actions of menu items in some versions of Microsoft Office where the items displayed in a given menu reflect the most recent set of items selected. This example is a particularly apt one of scaffolding changing inappropriately, as for many users it is disorienting and caused more problems than a non-changing menu set did.

In dynamic systems for daily life support the scaffolding can manifest as changing levels of detail in task support, based on use history (which is part of the user model) and on sensor data. For instance, if a set of prompts for a task was rapidly triggered through, faster than actually doing the steps, the system could contract those steps into a single step. Just as internalization of a sub-task can trigger scaffold retraction, so too, for an aging population with progressively decreasing cognitive abilities, must scaffolding automatically extend to provide well-fitted support.

2.4.1   CANONICAL PAPERS

The research on the subject of scaffolding comes primarily from educational domains, but significant work in other domains is presented in Part 3.

Wood, D., J.S. Bruner, and G. Ross (1976). "The role of tutoring in problem-solving." (Wood et al., 1976).

Wood, D. and D. Middleton (1975). "A study of assisted problem-solving." (Wood and Middleton, 1975).

2.4.2   AT DESIGN EXAMPLES

In MAPS trials, in some cases, as parts of a script become memorized and become part of an internal repertory of sub-scripts (see Figure 2.17), the longer set was replaced in a subsequent script with a single cue for that internal memorized set of instructions. For instance, a script for making cookies had 40 steps and one trial participant initially started the script by looking at the prompt, sometimes repeating the verbal instructions, and then following the instructions (i.e., remove this ingredient from the refrigerator, take the baking pan from the cabinet, turn on the oven…), when she came to the 10-step instruction sequence for actually mixing and placing the cookie dough she skipped through the steps with pauses between each prompting screen, and, at the end of the set, stopped and prepared the cookies on the baking pan. When she finished that set of steps she went back to the look, act, look loop until the end. The trial observer saw this and confirmed the observation by reviewing the log and saw that the set of steps were displayed with only seconds between them until the log came to the script step that went back to the 10 or so second span for looking and acting. In this case, the participant had made these cookies (the script had been made by a special education professional from a standard set of cooking scripts for people with cognitive disabilities) previously and had internalized the steps. In this case, the next time the script was given to the student the set of prompts for that sequence was replaced with two steps and she had no problem accomplishing the task.

In another task, an adult with intellectual disabilities was being guided in folding clothes from the dryer. Interestingly, this person held down a job in a local gym, was able to cook and take a bus to work independently, and had a personal bank account, but seemed to never be able to succeed in folding and storing his clothes from the dryer properly. The segmented steps of the task were sufficient for the talks with one exception—he was not able to master the folding and hanging of his pants properly, so his parents rewrote the script to expand the pants folding prompt into many steps and he was successful. As his use continued he became better at pants folding and the steps were contracted as in the example above. In MAPS, the expansion and contraction of the scripts scaffolding were done manually based on logs and observations, but it is not difficult to envision an automatic expansion and contraction based on logs when new scripts were loaded, and further to have the dynamic prompting happen in the process triggered by task performance in real time. One way to capture this might be to monitor the logs and react to cases where the end-user rocked back and forth over steps in the script and have the prompting system dynamically expand the sub-set of prompts to give more detailed guidance.

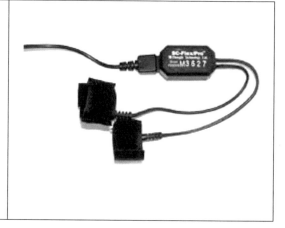

Figure 2.19: Components of Mohamed's scaffolded game system. From Mohamad (2005)

Yehya Mohamed's work on dynamically expanding and contracting a task level of difficulty based on biometric information is an example of dynamic scaffolding (Tebarth et al., 2000; Velasco et al., 2004). His adaptive interface is an excellent example of dynamically extending and contracting scaffolding (Mohamad, 2005) (see Figure 2.19). He evaluated existing adaptive systems and developed his by analyzing the arousal states of the user with non-invasive biofeedback sensors. The dissertation project provided a reliable means to recognize the arousal states of the user that complements research attempts in this area that interpreted arousal states by other external signals of the human body, like gestures, speech intonation, blood pressure, skin conductivity, etc. His system used a combination of sensors and a set of effective algorithms to infer the arousal state of the user (Figure 2.20). This information was used to modify the behavior and the interfaces of the application (Figure 2.21). In particular, a therapeutic system was developed where Interface Agents—which themselves can communicate through mimic, speech, and affective expressions with the user can modify their behavior according to the arousal status of the user. His system uses the skin conductivity sensor as the input device to measure the user's arousal states. By monitoring this biometric information and estimating emotional levels he is able to make the elements in the pedagogical interface more or less difficult, thus by keeping the interface of system adapted to the optimum flow of the user he maximized the pedagogical efficacy of the system. He does this by having the system adapt the task difficulty to the maximum limit of the child's abilities by analyzing continuously the child's performance. When the system determines that the child's perception of the difficulty of the task becomes too low, it adjusts the task to be more challenging; and when it is too hard, it makes the task harder. In this way it keeps the level of challenge optimal for the learning process.

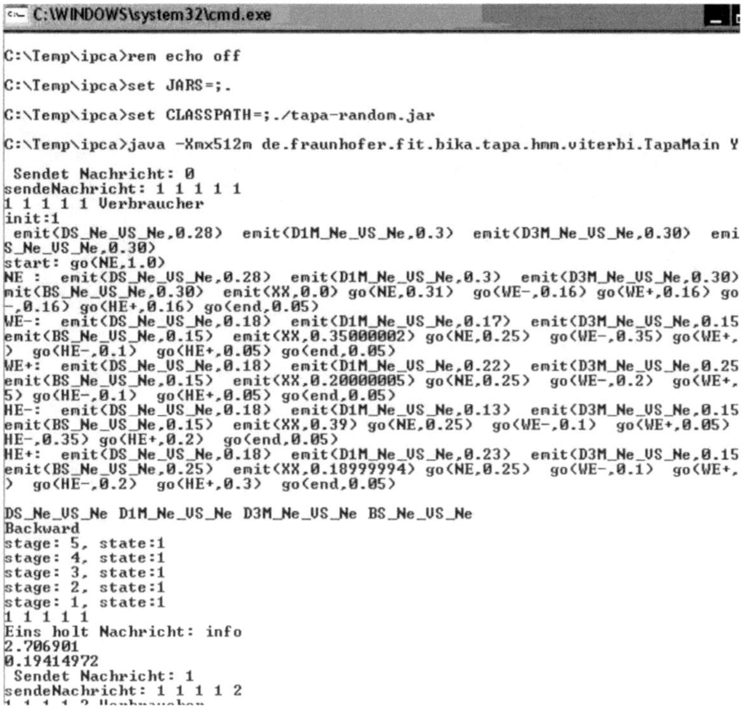

Figure 2.20: Raw sensor data and classifier. From Mohamad (2005).

Figure 2.21: **TAPA** game interface. Fom Mohamad (2005).

In conclusion, dynamic expansion and contraction of guidance and increasing the difficulty and ease of effort presented to the user (isomorphs of the same problem), can be a good way of fitting the tool to the model of the end-user as internalized scripts and skills change.

2.4.3   CONCLUSION

Scaffolding has tremendous promise for AT/IA, but with few exceptions (Mohamad, 2005) existing implementations require manual intervention. Truly dynamic scaffolding will require advancement in contextual sensoring and inference, which is only available non-trivially in the laboratory or in research projects. Dynamic versions will be based on estar user modeling (a user model that reflects, and keeps this as history, the current actions and context of the end-user) (see Section 4.4) and inference machine learning systems. Perhaps the best approach is to start with the low-hanging fruit (see Section 4.2) that computationally emulates manual scaffolding extension and retraction. An exciting time to be a researcher!

2.5   SITUATED ACTION/COGNITION

Short Definition: Situated action "stresses the knowledge ability of actors and how they use common-sense practices/procedures to produce, analyze and make sense of one another's actions and their local or situated circumstances" (Doerry, 1995). Simply, situated action is how people act in a situation (Suchman, 1987). Sometimes referred to as situated cognition, the main point here is that actions in performing a task are often based on the actual situation at hand as well as any supporting guidance that is previously prepared.

Longer Description: Lucy Suchman, in her canonical exposition of this concept, said the following (Suchman, 1987).

> *Rather than attempting to abstract action away from its circumstances and represent it as a rational plan, the approach is to study how people use their circumstances to achieve intelligent action. Rather than build a theory of action out of a theory of plans, the aim is to investigate how people produce and find evidence for plans in the course of situated action. More generally, rather than subsume the details of action under the study of plans, plans are subsumed by the larger problem of situated action.*

The critical point here is that, in the real world, plans and guidance are almost always not literally sufficient to accomplish the task at hand. Closely examining any sufficiently complex set of actions, it becomes clear that the tacit knowledge of the world and domain needs to be added to the set of instructions to be successful. More importantly, mis-estimating the background knowledge held by the performer can sabotage the activity. We have all seen this in an attempt to guide elderly relatives to fix a computer problem that to a technologist is trivial, with a common result that the elder feels that there is "something wrong with them" in trying to act on the instructions. This particular experience can lead to abandonment of the use of the technology entirely.

Therefore, the first important point is to correctly identify the level of tacit and implicit background knowledge the targeted end-user may bring to the use of the system. Furthermore, what seems obvious to the designer may be mysterious to the end user not only because of a murky interface but because of a lack of a useful model of what is going on in the background of the task Note that I used the term "useful" rather than "accurate" in describing the model that the user projects on the technological system; it is not necessary that a user understand the actual details of the system, only that their understanding maps enough to the system for their purposes. An example is that of modern file systems in personal computers, in order to copy a file from one folder directly to another there is no need to understand the representation that a windowing OS provides nor FAT tables or Inodes.

Second, what looking at a task through the situated action lens leads to is thinking about error trapping and mitigation (see Section 2.2), and enumeration of possible points in the act that may need provided help in the right way at the right time. *Tool tips*, while not widely used in AT, are a good example of providing contextually appropriate help at the right time.

Finally the insights of situated action provide a theoretical basis for having an expectation of the inevitable failure of any set of instructions or prompts due to the intersection of environment, infrastructure, user, and artifacts. Armed with this insight, a designer of AT tools for intelligence augmentation is, to some extent, protected from making naive assumptions as the basis of their design process. More about this is discussed in Section 2.2.

The above discussion takes a specialized perspective on the much more broad insights that Suchman and other researchers present. However, the core insight of the inevitable inability of any set of guidance tools in solely supporting task completion has been a constant guide in whatever success I may have had in projects and systems.

2.5.1   CANONICAL PAPER

Suchman, L.A. (1987). *Plans and Situated Actions: The Problem of Human-Machine Communication.* (Suchman, 1987)

2.5.2   AT EXAMPLES

Situated action/cognition is not really a basis for design of AT/IA, as discussed above. However, the design process of the last (and first) kilometre guidance in ASSISTANT (see Section 4.2) was a result of putting together situated action with the process of design for failure (see Section 2.2). The designers knew that, for urban pedestrian navigation, GPS was not a reliable guide. Yet, the project was committed to provide support for this critical part of the elder's journey. So taking those parts of urban GPS that were reliable (see Section 2.2 for details), the system recognized when there were breakdowns in the process (being careful to avoid type one errors) and provided support for end-users to (1) know that something was wrong and (2) give them the most reliable information available to remedy the error.

2.5.3   CONCLUSION

The importance of the perspective of situated action is to make the designer humble. Suchman's gift to AT/IA is that you will never be able to simply support task completion in a bubble no matter how sophisticated you get. And it works both ways— determining what the user is trying to do on the basis of observed behavior (basically sensor inputs) is never trivial. For me, reading her work was, in my humble opinion, something like Gödel's theorem. There is not a closed system solution for many of our problems, at least those in the natural world (i.e., with the exception of systems that are self-contained, like Coach).

2.6   SOCIO-TECHNICAL ENVIRONMENTS

Short Definition: Socio-technical environments (STE) or sociotechnical systems (STS) are a way of looking at technology, workplace practices, and end-users (as well as all stakeholders) as an integrated whole. By using the whole system or environment as a unit for analysis, the fitness and efficacy of the system is improved and the dignity and intelligence of the human elements can be supported.

Longer Description: The development of the socio-technical approach to technology system design is well documented (Coakes et al., 1999; Mumford, 2009). Looking at technology and its use in STEs came out of the Tavsitock group in post-war England. STEs are composed of two basic elements: the technical part and the social part. The technical subsystem comprises the devices, tools, and techniques needed to transform inputs into outputs in a way that enhances the economic performance of the organization. The social system comprises the employees (at all levels) and the knowledge, skills, attitudes, values and needs they bring to the work environment; additionally, the socio element encompasses the relations between the social roles, including authority and legal dimensions. By looking at the world of work as a whole system and placing equal emphasis on productivity and ethics, early advocates of the socio-technical method were able to elicit detail and dynamics of the elements of the system that previous (Taylor, 1967) oriented industrial engineers had missed. Furthermore, by utilizing a psychological perspective without class bias they were able analyze the system with a raw honesty that the earlier industrial researchers missed owing to their allegiance to using only an emic perspective (the story that the management tells about the institution's operation) in contrast to the STEs more etic examination style (the actual behavior of the system as perceived by an outsider).

By explicitly concerning themselves with the human values, STE practitioners are included as part of evaluation of a work environment. Also by concerning themselves with the more operational aspects of work systems, such as proper location of boundaries where knowledge is shared and the need for minimal yet adequate design of the process advocates of the STE approach, argues that ethical concerns and efficiency were not mutually exclusive goals in work process design.

Implicit in this approach is a belief in the importance of humanistic principles, and that one of the main tasks of the system designer is to enhance the quality of working life and the job satisfaction of the employee. In turn, they felt that the achievement of these objectives will enhance productivity and yield added value to the organization. At the root, the contribution of socio-technical design was to place the rights and needs of the employee as high a priority as those of the non-human parts of the system. What resulted were diagrams like Figure 3.3, where the employees are equal parts of the study rather than instruments of connection or as another machine to transform raw material.

Classical dimensions of concern to STE analysis are job enrichment, job enlargement and rotation, motivation, process improvement, task analysis, and work design (PTG Global, 2009). Frey (2009) posits four aspects of STEs (or in his case he calls them socio-technical-systems (STS): (1) STE components are interrelated and interact so that a change in one component often produces changes in the other components and in the system as a whole; (2) STE have different components (i.e., hardware, surroundings, software, groups and roles), which interact with one another; (3) socio-technical systems embody values, and this embodiment can be identified in specific components in the system; and (4) STEs change, and this change, referred to as a trajectory, reveals internal relationships and must be accommodated and perhaps canalized for the system to succeed in all dimensions of measurement (i.e., efficiency and satisfaction). The elements of his analysis are: hardware, software, physical surroundings, people (groups and roles), procedures, laws, and data and data structures

The developers of "classical" socio-technical theory created a number of frameworks and tools for analyzing and designing socio-technical systems. Enid Mumford, one of the seminal workers in socio-technical analysis, developed an approach to system analysis and design called the ETHICS method (Mumford, 2009). ETHICS consists of a series of steps that progressively involve the various stakeholders in the design and adoption of a new system.

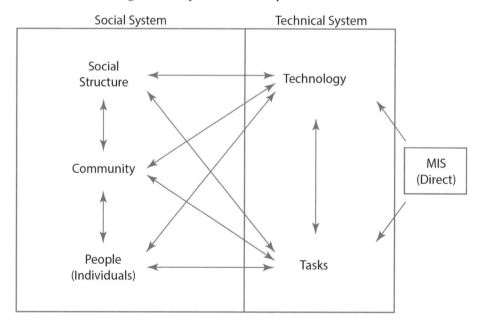

Figure 2.22: Elements of a socio-technical system.

Of particular use to analysis of large and seemingly amorphous systems is the STE Network Model approach. In this approach the researcher uses graph modeling methodology to map out elements and known relations between them, and in the process may find existing tacit connections and missing, but necessary for optimal performance, functional elements, and connections (Scott, 2000).

Enid Mumford (2003), in her book *Redesigning Human Systems*, presents several personal cases of STE analysis and STE-supported design. Following the guideline in the STE approach of understanding current work practices before attempting to re-engineer the system, Mumford closely observed the process getting to know the participants, environment, and procedures. Her techniques, in one case of getting a job of a canteen worker in a loading dock area, while not formally using ethnographic protocols, approximated classical ethnographic work. Ethnographic participant observation (Bernard, 2000, 2002) becomes highly relevant in understanding both the present process and the iterative design and implementation of new processes. Mumford's approach to respecting and learning from the current practitioners of a work practice and environment is echoed in a quote from Mao Tse Tung that she presents in her book (Mumford, 2003):

> *Go to the practical people and learn from them: then synthesise their experience into principles and theories; and then return to the practical people and call upon them to put these principles and methods into practice so as to solve their problems and achieve freedom and happiness.*

Working on AT systems with an STE perspective is an instance of a wicked problem (see Section 2.8), in that the final form of the solution is typically not specifiable at the time of the initial design. This is because complex computer systems that are designed for interaction with people (in contrast to, say, rocket guidance software) are only properly studied in place. Since the system, its context, and the human users form a socio-technical environment, this is not ready to be studied until at least a rudimentary environment is in existence. Because the STE is comprised of the system itself and the members (actors/stakeholders) of the system and the interactions and the changes in work practices, the introduction of the computational artifacts induces an analysis of the problem that is complex and time consuming. Compounding the problem is that they often have multiple stakeholders (some of which are very tacit, e.g., the legal system). An example of the consequences of not studying the whole range of stakeholders and the changes that a new system induce in work practices was the result of a complex, very user-friendly, system. It was carefully crafted with user input to work with minimal effort in a system that enabled special education teachers to share best practices with AT and approaches in the classroom. Upon test installation, the school administrators demanded the removal of features that made the system unusable by the very end-users, the teachers, who needed these fucntions the most. This was done by the school administration who, under HIPAA privacy legal constraints, were advised by legal staff to remove any identifying information about the students, the very information (even when anonymized) that the system was designed to share (dePaula, 2004). Interjection of computational artifacts changes the environment's work practices, which may affect many more shareholders than expected.

Good AT design is then best approached from the STE perspective. The creation of AT for people with cognitive disabilities is particularly an STE issue due to the complexities of relationship and invasiveness of the technology. Following on this is the question of how best the STE approach can be formalized in AT design. Frameworks, such as the ETHICS method (Mumford, 2009) and the decomposition into facets as presented by Frey (2009), would be good places to start. Most useful is the emphasis on ethnographic study forming the basis for understanding the potential and requirements of the situation. An explicit drawing out of stakeholders' roles and investments in the whole system is illustrated in several ethnographic studies in Mumford's works. Illustrating a similar approach is MAPS inclusion of the caregiver's role and interface design, which naturally flows out of approaching the problem from an STE perspective, a concern that is often missing in other high-functioning but low-adoption systems.

Developing an explicit checklist set of heuristics to incorporate STE perspective in AT design is another approach that may roll back the tremendous problem of abandonment. Finally, STE's systemic approach of acknowledging the dynamic interaction between user, artifact, environment, and tasks is critical for good AT design.

2.6.1   CANONICAL PAPER

Mumford, E. (1987). Sociotechnical systems design: Evolving theory and practice, in computers and democracy, (Mumford, 1987)

2.6.2   AT EXAMPLES

AT Design and Use Example: MAPS

Ethnographic Design: Because MAPS was intended to be used by a population with abilities and needs very different from the designer and programmer on the project, the initial steps of the design process were to gain understanding of the end-users and their tasks. Various ethnographic approaches were used to support this. Initially, spent time was spent with professional caregivers and teachers who specialized in working with young adults with cognitive disabilities. The object was to understand the young adult with cognitive disabilities and their caregivers, both in day-to-day life and in contexts where they might use MAPS. By doing ethnographic participant observation (LeCompte and Schensul, 1999) the MAPS designers gained an understanding of the young adult with cognitive disabilities abilities and preferences that informed the prompter design. The observation process also provided example tasks that MAPS could support, as well as provide details about the environment the tasks would be accomplished in. By observing and analyzing the observations of the interactions between the end users and caregivers in daily life the relationship dynamics were exposed in a way the simply gathering formal requirements for a software project could not do (LeCompte and Schensul, 1999).

In practice, what this meant was spending about 20 contact hours with each dyad (i.e., pairs of end-users and caregivers; see Section 3.1), 10–15 hours with just the young adult with cognitive disabilities, a lesser amount with just the caregiver, and the balance with the caregiver and the young adult with cognitive disabilities. The goal was to learn how each of them interacts with the world and with each other. One way this paid off was the discovery that sometimes the verbal prompts should not be recorded by the caregiver, especially so in the dyad consisting of a mother and 16-year-old daughter who had typical adolescent power issues; this highlighted the fact that although these end-users were developmentally disabled they would often go through the same changes in relationships within their family as their non-developmentally disabled siblings. Spending time with the caregiver gave the designer a better idea of the level of PC expertise that would be available and also helped to give the caregiver realistic expectations for the MAPS system. Most importantly, participant observation gave a window into how task accomplishment was currently being supported and provided a reference against which MAPS could be measured.

Having produced a prototype, the second part of the MAPS project was to study the initial use, iterative design changes, and final adoption of the system in a real world context. With each dyad there was a process of generation and use increasingly challenging set of scripts. The content and environment of the scripts was typically from simplest to most complex:

- controlled environment (e.g., a housekeeping chore), in which neither the task nor the environment is dynamic and the environment is familiar;

- less controlled script (e.g., cooking), in which the task doesn't change and the environment is dynamic but familiar; and

- least controlled script (e.g., shopping), in which the task and the environment are unfamiliar and the environment changes.

The first script was also used to familiarize the caregiver with the script design and composition process and the person with cognitive disabilities with the use of the prompter, its controls, and how to follow a script. The final two were designed to simulate, in a real context, increasingly autonomous use of the system. The three-script process allowed changes to be introduced into the system to make it better fit the needs of the task and user. One example was the addition of multiple scripts that could be chosen by the young adult with cognitive disabilities depending on the state of the tasks. Another dyad changed the script between iterations several times to accommodate learned sub-tasks that were initially cued in detailed prompts and now could be called up with a single prompt.

The most obvious connection between the STE approach and successful AT design is the equal focus on the mechanical/technical aspects and the human aspects of the whole system. This section will map some of the more specific topics and schemas from the STE literature and AT design.

Mapping the specific parts of the ETHICS method (Mumford, 2009) to the process of designing MAPS shows a congruence that might make for a good framework for AT design in general, as well as describing the process of choosing and adopting a new piece of AT. Starting with ETHICS first step (why change?), which corresponds to the AT professional determining not whether a change is needed, but what form the intervention will take, the disability itself is the motivator for wanting to change. ETHICS next step (system boundaries) corresponds to the caregiver and AT expert determining what the person with cognitive disabilities might be able to do with technological assistance, and what they should not attempt to do. ETHICS step 3 (description of existing system), which in the STE process describes the procedure of learning the current system (often from the inside out), corresponds to the AT designer studying the work practices and day-to-day details of the person with cognitive disabilities and the dyad's life, as well as the occupational therapist or special education professional detailing the end-users existing problem solving personal adaptations. All too often high-functioning AT in the form of a cognitive prosthetic fails in functionality and/or adoption because the designers naively felt that they understood the needs of the end users, an insight best gained through personal involvement with the person with cognitive disabilities. Similarly, both the AT designer and the OT or AT expert that recommends and guides the end-user in obtaining AT need to perform the definition of key tasks which includes both teaching task segmentation and the use of a prompter.

The future step analysis corresponds to the AT designer explicitly making the AT re-configurable to support future needs, in some cases providing less support to the end-user—as in the case of acquisition of skills—and in other cases providing more support for greater levels of disability. This step also corresponds to the notion of an STE system accommodating system change, the trajectory in STS terms (see Socio-Technical Systems in Professional Decision Making Module (Frey, 2009)). This lack of support for co-evolution causes much of the abandonment (Phillips and Zhao, 1993) of AT tools. Caregivers, who have the most intimate knowledge of the client, need to become the "programmer/end-user developer" of the application for that person by creating the needed scripts. Similarly, the other steps in ETHICS can be mapped to stages in AT design or adoption.

Successful AT design and STE analysis are both based on worker (end-user) participation and management (caregiver and AT professionals) support. It is only with a base of both of these that the use of, and appropriate AT aid, can continue past initial use. Suchman (1987) describes the actual trajectory of attaining a goal as being heavily dependent on situated action—on accommodating changes in the context and requirements to attain a given goal. Therefore, both STE systems and successful AT must be able to accommodate change in use, what Frey and others (Geels and Kemp, 2007; Frey, 2009) refer to as trajectory. For an STE to continue to be relevant and appropriate it must adapt to coordinated sets of changes within the socio-technical system. This coordinated series of changes in an STS is called a trajectory. Trajectory must fit situational needs if STNs can support/withstand change. It is also a quality of successful AT for people with cognitive disabilities such that it must reflect change as the system does not need to fit into "artificially" created (and thus stable) work practices but into the world as it is, which is constantly changing. In the case of MAPS, the ability of the caregiver to edit the MAPS scripts, enlarging them if there is greater need, compressing them in the case where sub-tasks become learned.

Decomposing AT into STE Grid

Frey (2009) uses a schema to analyze socio-technical systems that divides the system into seven components. These parts—hardware, software, physical surroundings, people (groups or roles), procedures, laws, data (and data structures)—give a good starting point for comparing different STEs and also for clarification of the interaction between the technical and social dimensions of an STE. Following is the decomposition of the MAPS system.

- Hardware: MAPS uses a PC for its MAPS-DE script-designing tool, feeding the script composition are recorded voice prompts and images collected by a digital recorder and camera, respectively. MAPS scripts are played on by the MAPS-PR on a PDA or smart phone that runs one of the mobile versions of the windows operating system.

- Software: The MAPS system software consists of the MAPS-PR script player and the MAPS-DE script editor. Optionally, in addition to these (and in support of them) are an image editor for the pictures illustrating the prompt, and an audio editor for the verbal prompts. Behind these are the Windows desktop and small device operating systems. Because one of the functions not disabled in the PDA was the MP3 player (to motivate retention of the PDA), the MP3 player application was also sometimes used. Additionally, some caregivers used a text editor (like MS Word) for preliminary script design. The scripts themselves were stored in a Sybase database on the PC and PDA, as well in as in a MAPS script template server that held pre-outlined typical scripts accessible in the Internet.

- Physical surroundings: MAPS was used in two kinds of environments. The MAPS-RP was used wherever the end-user was performing tasks with the aid of the scripts prompting. In the initial trials of MAPS these ranged from the end-users home to a school to employment (i.e., in a used clothing store). As well as the MAPS-PR being used in these spaces, the caregiver would photograph prompt visuals in them for script creation. The prompts were most often recoded in the home, or in the case of the job coach, the office of the caregiver. In the case where the MAPS-PR PDA was being used as a MP3 player, the location varied with the path of the end-user through the day.

- People: The list of people includes not just individuals (roles) but also groups of people (groups). These include the designers of the MAPS system and the end-user co-de- signers. Central to the socio-technical system are the end-user (also referred to as the MAPS-PR-users, a person with cognitive disabilities, and the client) and the day-to- day caregiver, who may be a family member or a professional caregiver, paid for by insurance, the family, or the state. Influencing the system at a remove are AT experts, special-education experts and teachers (in the case of a young adult with cognitive disabilities), insurance personnel, and state funding staff. At a further remove, but still very much affecting the system, are school administrators and employers. Finally, intimate influencers of the system are the end-user's immediate family and friends, as well as their peer groups (either in school or employment).

- Procedures: There are several kinds of procedures in MAPS, in setting up the system and in using it to guide task performance. Included in the tasks required to setup the system are task segmentation (i.e., breaking the task down into sections that are of the correct cognitive level), task rehearsal (i.e., performing the task yourself to ensure no tacit steps are left out), and script building. The construction of scripts (after the out- lining has been done) requires collecting photographs of the task with a digital camera and recoding the verbal prompts with a computer and mike. The caregiver must master the art of using the MAPS-DE using the provided tutorial. Script assembly requires using the MAPS-DE editor and the operating system to identify and insert script steps into the script database.

Next, the caregiver has to transfer the script to the MAPS-PR from the caregiver's PC. The end-user has to initially learn how to use the prompter by working with the caregiver and perhaps the MAPS design-support personnel. The young adult with cognitive disabilities is then ready to use the script on the MAPS-PR to accomplish the task, which is embedded in the larger set of ADL and IADL tasks that she can do without external support.

Finally, the caregiver has to review the script log to see if the script needs to have cer- tain steps collapsed into a trigger step (collapsing scaffolding) or expanded into several additional steps because the end-user found performing them too difficult (expanding scaffolding).

Laws, statutes, and regulations: The MAPS system is not impacted by laws and regulations except inasmuch as it's purchase can aided by state funding.

• Data and data structures: MAPS stores external wave files (for the recording prompts) and jpg files (for the prompt images) in the caregiver's PC. Completed scripts are stored in a Sybase database on the caregiver's PC and scripts on the MAPS-PR are stored on a mobile lightweight version of the Sybase database. Additionally, a Sybase database of template scripts is stored on a networked server, accessible through the Internet. Design documents used in creating scripts (i.e., task segmentation notes) may be stored in text documents. MAPS-PR stores a log produced of the use of a given script in a text file for later analysis.

Finally, in this comparison between AT design and adoption and traditional STE analysis (Table 2.3) there is in both a concern of studying work practices or in the case of AT, the activities of ADL and IADL, as they are. Therefore both disciplines use varying forms of ethnography and base their theoretical analysis on both foreground (worker/person with cognitive disabilities) and background (environments, rules, and technologies).

Table 2.3: Comparing industry and AT and STE

	Hardware	Software	Physical Surroundings	People (Groups and Roles)	Procedures	Laws	Data and Data Structures
AT Example (MAPS)	PC and PDA prompter	MAPS-DE and MAPS-PR	Caregiver's home, end-user's world	Client caregiver maps developer AT special-end experts	Creating and editing a script, using prompter to do a task, segmenting a task	Privacy and copyright laws	Scripts database, folder of images, folder of recorded prompts
Industry Example: Volvo (Mumford, 2003)	Assembly tools	Worker Protocols	Automobile factory	Workers, management, line supervisors, workers families, researchers	Assembling automobiles, updating work practices, updating team practices	Worker rights, economic realities, contracts	Assembly drawings, workorders and bills of materials, lean production artifacts

STE and AT design particularly impacts "workplace practices" in the sense that introduction of new AT technology may change a "stable" family dynamic in very unanticipated ways. The in-house special education consultant we had in our lab at L³D told us her experiences (Kintsch, 2002) with the introduction of AAC (assistive and augmentative communication devices) into a family that had worked for years to create a family dynamic that worked for them based on a member being the communication helper for the child with a fairly severe communication disability that required the family member to be the one that was the only way to support communication for the child with the disability within the family and with the outside world. This stable dynamic was threatened by the introduction of the support AT and she had seen several situations where the caregiver subtly sabotaged the new configuration, presumably threatened by the new and unfamiliar family dynamic. By being aware of all the stakeholders and anticipating the sorts of unintended changes AT introduction can cause, a painful cause for abandonment can be circumvented (Kintsch and dePaula, 2002). Similar to this situation is the use of AT in school and transitioning into the use of the same AT in the home environment.

2.6.3   CONCLUSION

Every introduction of technology is a socio-technical event. But every AT system does not necessarily benefit (e.g., in a cost/benefit way) from basing the design with STE. It's important to evaluate how much the work practices will change, how many "shadow" stakeholders are involved, and the intent of the AT system. For the systems that this book describes the STE approach is recommended, but this needs to be a decision not an assumption. Doing STE-based research properly is not trivial; it can be quite time, and resource, consuming.

2.7   UNIVERSE OF ONE

Short Definition: People who could most benefit from high-functioning AT are also often afflicted with co-morbidity, resulting in users that are not effectively generalizable as a design target.

Longer Description: *Abandonment Based on the "Universe of One."* People with cognitive disabilities represent a "universe of one" problem (Fischer, 2009): a solution for one person will rarely work for another. Accessing and addressing these unexpected variations in skills and needs, particularly with respect to creating task support, requires an intimate knowledge of the client that only caregivers can provide (Cole, 2006). The consequence of the high rate of device abandonment (Phillips and Zhao, 1993) is that the very population that could benefit most from technology pays for expensive devices that end up in the back of closets after a short time.

AT for cognitive disabilities (and to a lesser degree all AT) presents unique design challenges stemming from being in the intersection of AT and cognitive science. One aspect of this is that due to the distance between the experience of the designer and the end-user's systems, which are often inappropriate or ineffective in real context of use. In attempting to understand the needs of the user and the task to be performed, the system designer can take one of two naive approaches. One could label these two problems "I've got a cousin" and "I've got a theory" (Section 5.2).

Disabilities are often complex mixtures of separate needs and lacks. Medical papers often refer to this as co-morbidity (Jakovljević and Ostojić, 2013), a useful word as well as sounding like what it describes. Especially in the domains of people with cognitive disabilities these multipliers of barriers make design of support systems and extensions of existing abilities very difficult. One size never (or very rarely) fits all. So the CLever group discussed what we were doing was designing for a universe of one, in contrast to making a single system that would properly fit many.

Reduced intellectual ability is often combined with sensory impairments such as visual and hearing impairments; with motoric impairments making input to the device require special adaptations; and psychological/developmental impairments affecting comprehension of the system and its outputs as well as problems with losing or not properly caring for devices and becoming impatient with incomprehensible or delayed outputs.

Compounding this problem, it is typical for novices to this kind of population to rely on diagnosis as a basis for design, starting with problem specification. This is exactly what we in CLever did, initially, until we had the good fortune (and excellent planning by Gerhard Fischer) to hire a formally trained special education professional with many years' experience. What resulted was a focus on functional disability, which allowed more effective persona building and problem specification. One of advantages to focusing on functional needs, in contrast to diagnosis, is the curb cut[10] effect that produces benefits for unintended populations (e.g., curb cuts for wheelchairs and baby carriages), as well as benefiting those with temporary conditions that produced disabilities. Examples of this are that driving a car lowers the cognitive and motoric abilities of the end-user, and being confined to a wheelchair while recuperating is not fundamentally different than using one on a day-to-day basis.

Universe of one dilemma is one end of an axis and the other end is "truly one size fits all." All (interface) design problems fall on this axis. By designing too tightly to one person, others cannot benefit; by only designing generic solutions, often those who do not fall into the typical range are excluded. One solution to this dilemma is deep personalization (see Section 4.4). There is validity to the of repeated motto "design for all just good design in the first place," even in some cases for the problems discussed here, but unfortunately the seven-point slogans of the Design for All movement (Center for Universal Design, 2011) really don't cover many of the multi-layered and interactive needs of intelligence augmentation.

---

[10] The curb cut effect: technology intended to benefit people with disabilities sometimes wind up benefiting everyone. Curb cuts, which are intended for wheelchair users to be able to get onto sidewalks, help bicyclists, parents with strollers, delivery people, and a dozen other nondisabled groups. Similarly, closed captioning, which was originally meant to benefit deaf people, helps people who have trouble with auditory information processing, people who like talking during films, and people trying to watch TV in noisy bars.

2.7.1   CANONICAL PAPERS

There really is no canonical paper on this topic, but this is a good medical overview:

> Jakovljević, M. and L. Ostojić (2013). "Comorbidity and multimorbidity in medicine today: challenges and opportunities for bringing separated branches of medicine closer to each other." (Jakovljević and Ostojić, 2013).

Also, Cole's position paper in a CHI workshop is very useful:

> Cole, E. "Patient-centered design as a research strategy for cognitive prosthetics: Lessons learned from working with patients and clinicians for 2 decades." (Cole, 2006).

2.7.2   AT EXAMPLE

Deep personalization in ASSISTANT. The project assumed, from the initial requirements stage, that the target population would be quite varied. Notwithstanding the "bandwidth of need and ability," discussed in the Section 2.3, making the target group much smaller, within this group there was a fairly wide variation in what the focus groups expressed as needed features. One elderly woman said that she would be anxious having to hold the PND equipped smartphone out in public, fearing theft. So ASSISTANT provided a voice only guidance feature, allowing them to use a mono ear bud or a Bluetooth earpiece. Some of the elders had a slight tremor in touching the screen, so the personalization tab provided an option to specify ignoring multiple rapid taps, while allowing more agile users to interact more rapidly. Some had vision problems, so there was an option of inverting the screen; some wanted to contact emergency help immediately, some wanted to have the system contact their relatives, so this was configurable by them in the route editor. Many more options were possible, or none, if the user chose.

In doing research for the optimal representation on the screen for image/audio prompts for the MAPS system and in talking with professionals in this field (special education and rehabilitation experts) I ran into comments that the more challenged the person is the more difficulty they have in using icons and generic symbols of objects and actions. Not finding satisfactory answers in a broad literature review, spanning studies on AT, rehabilitation and cognitive psychology, our team did an experiment (discussed in detail in the Section 3.2). As a result, the MAPS system used actual images of the prompts (e.g., vegetables bin at the real store where the user was instructed to pick a head of lettuce, the familiar cashier at the store the end-user and her mother went to regularly). To do this, a much more complex script editor was provided to the caregiver to assemble scripts to use photos that were for her alone. The resultant script was then tailored to the young lady and her mom. If you look at the resultant script as an application, then the problem of universe of one was accommodated. Interestingly enough, the mother in this case had me record all the prompts to accommodate her teenage daughter's resistance to being told what to do, something that was commonly unexpected on my part. Autistic teenagers, even not-so-high functioning ones, are still teenagers with all the emotions of any teenager.

2.7.3   CONCLUSION

Design for one is not as much particularly a design framework as a cautionary perspective prescription. Certainly deep personalization is part of the answer, as is a clear design problem statement. Bounding the problem to specific functional disabilities helps a lot. Best is interaction and study of the end-user and stakeholder population. These are sometimes quite different activities; when doing ethnographic research for the MAPS project, about ¼ of the time I was with the young adults and caregivers was spent just hanging out rather than focusing on the task at hand, producing unexpected insights and a motivational "push" that greatly improved the system.

2.8   WICKED PROBLEMS

Short Definition: A wicked problem is a problem that is difficult or impossible to solve because of incomplete, contradictory, and changing requirements that are often initially difficult to recognize.

Longer Description: Ill-defined design and planning problems can be labeled "wicked" (Rittel and Webber, 1984; Simon, 1984) in contrast against the relatively "tame" problems of mathematics, chess, or puzzle solving. Wicked problems have incomplete, contradictory, and changing requirements, and solutions to them are often difficult to recognize as such due to the complex interdependencies. Typically, wicked problems have the following characteristics.

- The problem is not understood until after formulation of a solution.

- Stakeholders have radically different worldviews and different frames for understanding the problem.

- Constraints and resources to solve the problem change over time.

- The problem is never solved.

- Solutions to wicked problems are typically better, worse, or good enough (satisficing).

In the domain of developing AT as IA the wicked problem space requires the two primary symmetrical holders of knowledge, the client and caregiver, to both contribute to the solution—the caregivers contributing the finished scripts and the client contributing the existing internal scripts (see Section 4.1). Wicked problems are not statically solved; rather, the on-going solution is a process. An example of this is the plotting of a bus route through a residential neighborhood, where the trade-offs include local passengers, property owners, traffic managers, and urban planners. Additionally, the route as planned may be good for only several years or less.

Wicked problems are most often "solved" (here the notion of satisficing emerges) through group efforts. Satisficing solutions are not true or false, but better, worse, or good enough. Task support through computationally based multimedia prompting is such a problem. The starting-off point for a designed solution is to do a stakeholder analysis of the problem space (Overseas Development Administration, 1995). Several types of stakeholders are involved in prompted task-support:

1. End-users: those with cognitive disabilities who will use the system to support doing the task;

2. Key stakeholders: those who can significantly influence or who are key to the success of the activity (in this case caregivers and clients);

3. Primary stakeholders: people who are directly affected by the solution; in this case, parents, employers, group home staff; and

4. Secondary stakeholders: all others with an interest in the activity; in this case, members of state and federal organizations who concern themselves with AT, insurance companies, and HIPAA bureaucrats.

Having identified the stakeholders, the designer then can proceed with a better assurance that the system can be adopted as a solution that must satisfice each class of stakeholder. Additionally, this analysis identifies the (sometimes orthogonal) requirements and identifies the symmetry of ignorance members of the problem space.

Some of the characteristics of wicked problems are:

• incomplete,

• contradictory, and

• have dynamically changing requirements.

Understanding this provides both solution constraints and also, perhaps more importantly, a tool for identifying the presence of this type of problem. Typically, the actual parameters of a wicked problem are not understood until after formulation of an initial solution, thus prototypes (tested by stakeholders) and models become important tools in the process. Because stakeholders have radically different worldviews and different frames for understanding the problem, stakeholder analysis and broad participation of different stakeholders in requirements gathering, like focus groups, becomes quite important. Because constraints and resources to solve the problem change over time, the solution must be adaptable, sometimes radically so. A well-functioning example of a satisficing solution is the U.S. Constitution and its amendments.

The problem is never optimally solved but solutions to wicked problems are typically better, worse, or good enough, hence the motion of a satisficing solution, one that disappoints all equally (Simon, 1996).

Does this sound familiar? What to do? Explicitly involve all stakeholders; underdesign solutions; allow for the seed, evolve, reseed process (Fischer, 1998); design for change (see Section 4.3)? Just knowing that the problem is wicked helps.

2.8.1   CANONICAL PAPER

Rittel, H. and M.M. Webber (1984). Planning problems are wicked problems. (Rittel and Webber, 1984)

2.8.2   AT EXAMPLES

Interestingly, it is difficult to find examples in AT/IA that are not successful due to the tacit implementation of the wicked problem design approach. In AT, the notion of iterative design is standard, however the need to involve all stakeholders to produce a satisficing solution is not always obvious from the beginning. An example of this is the rollout failure of DePaula's innovative special educators collaboration system (DePaula, 2002) due to legal blocks (see discussion in Section 2.6)

2.8.3   CONCLUSION

Like distributed cognition (see Section 2.3), wicked problems are everywhere. It may be, in the domain of AT/ IA, that it is better to start assuming that the problem is wicked. Contemporary software engineering, in the form of the agile approach (Cockburn 2001) does, to some degree, acknowledge this truth in the conventional application world.

CHAPTER   3

Models

This chapter presents several ways of modeling the problem and how to use these modeling frameworks:

- Dyads

- Importance of representation

- Tools for living and tools for learning

3.1   DYADS

Short Definition: High-functioning AT often has two first-class users, the typical end-user (the actual person using the AT to accomplish actions) and a person in the role of a caregiver. The caregiver is needed to help configure and personalize the complex technology as well as monitor the efficacy and update the data/personalization of the application.

Longer Description: A unique aspect of software and systems for people with cognitive disabilities is that, while the focus is on the end-user, the person with cognitive disabilities, its design and evaluation must involve their caregiver. In fact, it may be taken as an axiom (Carmien, 2007; LoPresti et al., 2008) that every system is used by and must accommodate a dyad—the end user and a caregiver. In MAPS design the end user was described as a dyad, one person with two roles, the person with cognitive disabilities and a caregiver, requiring two interfaces using one set of data. Typically, the caregiver assists in the setting up and maintenance of assistive technology systems, as they often are too difficult for the person with cognitive disabilities to setup and keep up to date. Also pertinent and contributing to the success or failure of a design being adopted or abandoned are the lesser stakeholders discussed above. Often the motivation and goals of these different stakeholder roles can be divergent and even conflict (Carmien, 2011).

IT-based AT will often have two end-users. There are several reasons for this; one is that often the preparation of the underlying data is too complicated for typical end-users. Examples of this could be creating multimedia prompt sets for a MAPS like prompting system or programming a route for a guidance application; also, the user interface (in contrast to the content) may need to be customized using an interface too difficult for the end-user. Both of these are really deep customization issues (see Section 3.2). The other reason is that it is often useful and perhaps necessary to provide a monitoring or active help function to the application that will appropriately bring the caregiver into the situation; it is critical that it does not flag too many false positives as that might lead to abandonment.

The other user, the nominal "end-user," must have an interface and set of functionalities that is appropriate to their needs. Interestingly enough, the design and implementation this part of the system is often an easier task. As a result, these types of interfaces have one database or set of static data and two quite different interfaces. In the case of Fidemaid (Carmien, 2009) (see Figure 3.2), a prototype system designed to support independence in elders in their financial activities, the caregiver interface allowed trusteed familiar caregivers, with the permission of the end-user, to monitor and set alarms for suspect transactions.

3.1.1   CANONICAL PAPER

Carmien, S. and A. Kintsch (2006). Dual user interface design as key to adoption for computationally complex assistive sechnology. (Carmien and Kintsch, 2006)

3.1.2   AT EXAMPLES

The MAPS system (Carmien, 2004b) is a very clear example of the dyad approach (Figure 3.1). The end-user interface was basically a multimedia player on a PDA. This was in 2004; currently, the platform would be a smartphone. The end-user (a young adult with cognitive disabilities) can start and step through uploaded scripts to complete a task. One version of the PDA prompter allowed the user to choose one of several sub-scripts at certain points in the task.

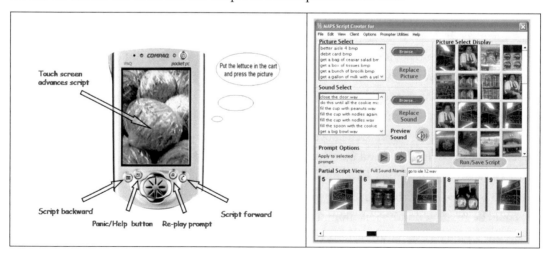

Figure 3.1: Left: the PDA prompter and controls; right: the caregiver's script editor. From Carmien (2006b).

The primary caregiver interface supported non-commuter experts creating, editing, and reusing scripts, which are sets of sequential multimedia (pairs of an image and a verbal instruction). The caregiver interface allowed the user to load pre-segmented task templates and populate the steps with their own images and recordings of prompt text. The caregiver could store and retrieve their own photos and recordings to use (and reuse) in script composition and editing. It also supported previewing the complete script.

A secondary interface for the caregiver was implemented in a follow-up project to MAPS, LifeLine (Gorman, 2005), that used tests imbedded in the uploaded scripts that tracked the performance of the task by transmitting the status of the script to a server and under certain conditions could send a SMS to the caregiver's cell phone.

The Fidemaid system (Carmien, 2009) was a research prototype to support elders in personal finance management. It was motivated to support elders living independently as long as possible and of the primary reasons for making the decision whether or not to move from their own home to supported living are health and medication issues and financial self management ability, as well as successful accomplishment of ADLs and IADLs (see Figure 3.2). The concept was to give them tools to easily see the status of their personal finances, both at the moment and over a span of time extending into the immediate future. Another support provided was a way to make decisions about major expenses by providing personalized comparisons between not making the purchase and the changes in discretional income that making the purchase would require. This prototype was based on research that discovered a lessening ability to make comparisons by casting both conditions into similar representations (Mather, 2006; Karlawish, 2008)

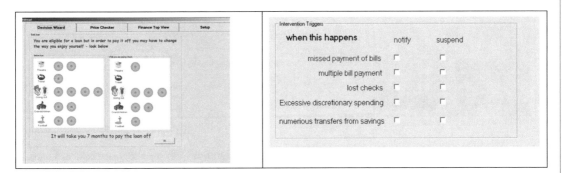

Figure 3.2: Fidemaid's (left) end-user's interface and (right) caregiver's intervention triggers. From Carmien (2009).

In Fidemaid's case the part of the system supporting dyads was the intervention trigger pane of the application setup. This allowed the end-user and caregiver to decide when and how to notify a *familiar* (Spanish for family based rather than professional caregiver) about potential problematic situations and incidents.

3.1.3   THE ASSISTANT SYSTEM

The ASSISTANT system, as discussed in various sections to follow, was designed so that, should the route planning be too complex for the end-user, a caregiver could do the planning for them. To do this the system provided a role of a "caregiver," or in the Spanish localized version the "cuidador," that allowed the end-user to delegate to a chosen, trusted person the ability to plan and send routes to the PND (see Figure 3.3). The system enabled multiple end-users for a caregiver and multiple caregivers for an end-user. This allowed the end-user to provide access to their data in a controlled fashion, ensuring the privacy and trustworthiness of the process.

Figure 3.3: Left, the end-user's PND; right the caregiver's assignment tab on the route editor. From Assistant Project (2015).

Not all support systems will require this two-interface approach, but often designers try and fit all the system's functionality into one interface. In both ASSISTANT and MAPS requests were made to put both the script/route design into the handheld device; issues with screen size, information complexity, and network robustness lead to the dual interface approach. Putting all the functionality into one interface may lead to an end-user user interface that is too complex to be used by the end-user, especially with respect to navigation within the system. The dual interface approach should, in my experience, be always at least considered, especially in the requirements gathering and initial design. This decision should be supported by all stakeholders, not just the technologist, and especially experts, clinicians, and family.

3.1.4   CONCLUSION

Not all AT systems that are IA based depend on the dyad approach, but many do. One important take-away from this section is for the designer to ask herself, "What part do caregivers or other intimate stakeholders take in the work practices that this system will require?" Even if the intertwinement is not as fundamental as in the above examples, the answer to the question will always improve the interface and functionality of the final system.

3.2   THE IMPORTANCE OF REPRESENTATION

Short Definition: Representation is the translation of components or strategies in problem solving to make the problem more tractable. "Efforts to solve a problem must be preceded by efforts to understand it" (Simon, 1996).

Longer Description: Solving a problem may simply mean representing it so as to make the solution transparent. Here is an example, a two-person game.

1. Take the numbers from 1 to 9 (1, 2, 3, 4, 5, 6, 7, 8, 9).

2. Players alternate and take one of the numbers.

3. The player who can add exactly three numbers in her/his possession to equal 15 wins.

Nickerson discusses this problem in his book, *Mathematical Reasoning: Patterns, Problems, Conjectures, and Proofs* (Nickerson, 2010):

> Sometimes the right representation can greatly reduce the amount of cognitive effort that solving a problem requires. A representation can, for example, transform what is otherwise a difficult cognitive problem into a problem, the solution of which can be obtained on a perceptual basis. A compelling illustration of this fact has been provided by Perkins (2000). Consider a two-person game in which the players alternately select a number between 1 and 9 with the single constraint that one cannot select a number that has already been selected by either player. The objective of each player is to be the first to select three numbers that sum to 15 (not necessarily in three consecutive plays). A little experimentation will convince one that this is not a trivially easy game to play. One must keep in mind not only the digits one has already selected and their sum, but also the digits one's opponent has picked and their running sum. Suppose, for example, that

one has already selected 7 and 2 and it is one's turn to play. One would like to pick 6, to bring the sum to 15, but one's opponent has already selected 6 along with 4. So, inasmuch as one cannot win on this play, the best one can do is to select 5, thereby blocking one's opponent from winning on the next play. In short, to play this game, one must carry quite a bit of information along in one's head as the game proceeds.

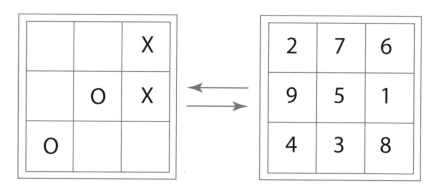

Figure 3.4: Tic-Tac-Toe → The same game based on Nickerson (2010).

One could reduce the memory load of this game, of course, by writing down the digits 1 to 9 and crossing them off one by one as they are selected. And one could also note on paper the current sum of one's own already selected digits and that of those selected by one's opponent. Better yet, as Perkins points out, the game can be represented by a "magic square"—a 3 × 3 matrix in which the numbers in each row, each column, and both diagonals add to 15. With this representation, the numbers game is transformed into tic–tac–toe. The player need only select numbers that will complete a row, column, or diagonal, while blocking one's opponent from doing so. There is no need now to remember selected digits (one simply crosses them out on the matrix as they are selected) and no need to keep track of running sums.

If you give this task (with timed turns) to a graduate student and a 12-year old, with the graduate student doing the number version and the 12-year old the Tic-Tac-Toe isomorph (Figure 3.4), the graduate student often doesn't have a chance.

Other representational "tricks" include this tree (Figure 3.5) to decode Morse code without memorizing the system.

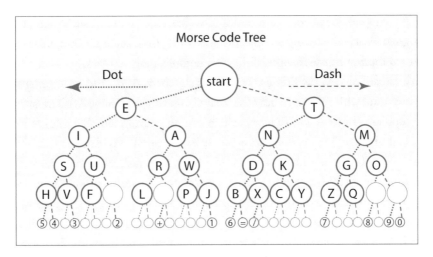

Figure 3.5: Morse Code decipher tree.

Another representational system leveraging everyday metal capacity is the Memory Palace (Yates, 1966). The memory palace is an age-old system, dating from Roman times through the Renaissance and featured in the current BBC television version of the Sherlock Holnes mysteries, for storing almost unlimited amounts of information and retrieving it easily (Ericsson, 2003) via a visualized house or location.

3.2.1   CANONICAL PAPER

Simon, H.A. (1996). *The Sciences of the Artificial.* (Simon, 1996)

3.2.2   AT EXAMPLE

The MAPS system used images to prompt steps in a script on a hand-held PDA (e.g., HP IPAQ). At first the intention was to use icons to represent objects and actions for the young adult with cognitive disabilities to be guided with. However, learning from the special education consultant in the L3D lab (Kintsch, 2002), and following the lead of the AbleLink commercial product (AbleLink, 2002), real and specific pictures were considered as better for recognition by the person with cognitive disabilities. A research literature search did not produce a tested confirmation of this assumption so the MAPS team decided to perform an experiment (Carmien and Wohldmann, 2008) to test this hypothesis. What follows is a summary of the paper referenced above.

In the process of developing MAPS, it became clear that the type of image displayed on the screen of the hand-held computer could affect strongly the success of the attempted task. The images used by computationally based augmentative and alternative communication systems (AAC) (Beukelman and Mirenda, 1998) and prompting task support systems can be a weak link in providing AT support for people with disabilities that require their use. When an image is not instantly recognizable as referencing an object in the world, then there are immediate consequences to the person attempting to use it successfully to accomplish a given task.

The two classes of systems that use images as supports, AAC and prompting support, have similar goals but differ with respect to how the image is used. Specifically, AAC images are intended to replace or supplant communicative ideas, which can range from simply pointing to an object in the world ("a loaf of bread") to encompassing an entire concept ("buy a movie ticket") (Beukelman and Mirenda, 1998). Computationally supported multimedia task support systems (Davies et al., 2002), on the other hand, have the ability to use a recording of a verbal prompt to provide the "verb" required to perform the task. Because of this difference, the emphasis of the present study was to determine the type of image necessary for successful identification of an object in the world.

The studies reviewed supported the notion that picture matching with objects is mastered before the use of images to communicate, and the matching skill is separate from the communicative skill. Several studies support the position that images out of context or, more properly in no context (on a white background), are more conductive to object matching. Other studies emphasized that picture recognition is a learned skill (as well as being developmentally based) and in support of this discussed various studies of different cultures approach to object representation and matching.

A common "rule of thumb" used by assistive technologists is that in order to obtain a match between a representation and the real-world object being depicted by the representation, the image fidelity must be inversely related to the level of cognitive ability. That is, lower abilities result in a greater need for fidelity of the image (e.g., Snell, 1987). Here the term "image fidelity" refers to a hierarchy of representations, from its actual physical form, to a model or photograph of the actual object, to a photograph of a specific brand of object, and finally to written words. It is not identical to iconicity or guessability, as the iconicity of a representation is a relationship between the observer, the representation, and the "target" object (e.g., Wilkinson and McIlvane, 2002). While research has pointed out that the iconicity of a representation is an idiosyncratic phenomenon, at least in specific instances, the authors posit that the need for higher fidelity is only loosely connected with cognitive ability—varying not only from person to person but also possibly from culture to culture (Huer, 2000).

The trials were designed to determine how the type of representation (icons, photos of objects in context, photos of objects in isolation) displayed on a hand-held computer affected recognition performance in young adults with cognitive disabilities. Participants were required to match an object displayed on the computer to one of three pictures projected onto a screen. The experiment was designed to test the opinion widely held by occupational therapists and special education professionals that there is an inverse relationship between cognitive ability and the required fidelity of a representation for a successful match between a representation and an external object. Despite their widespread use in most learning tools developed for people with cognitive disabilities, our results suggest that icons are poor substitutes for realistic representations.

Fifteen high-school students (9 males, 6 females, m = 15.8 years old) from the special education classes in the Boulder Valley School District were invited to participate in this experiment. Informed consent from both the student and his or her legal guardian was obtained. Students were selected on the basis of their IQ scores and their ability to use the MAPS script guidance system, as well as on their need for such a mnemonic and executive cognitive orthotic to accomplish activities of daily living (ADLs) independently. There was a wide range of cognitive disabilities, including, but not limited to, autism, Down syndrome, alcohol-related neurological disorders, and fetal alcohol syndrome. IQs (using the WAIS scale) ranged from 43–85 (m = 58.1). In addition, 15 age-matched high-school students (9 males, 6 females, m = 15.9 years old) with no cognitive disabilities were selected from classes in the Boulder Valley School District.

A laptop computer was to run a program written using Visual Basic V6 and Embedded Visual Basic. A projector displayed three target images onto a white screen. The computer was connected to a wireless router that established a local area network (LAN), which was connected to a hand-held Compaq IPAC model 3670 computer, with a 2¼ × 4 inch screen. There were three large push buttons (6 inches in diameter), in three colors (yellow, red, and green), attached to the laptop.

Details of Experiment

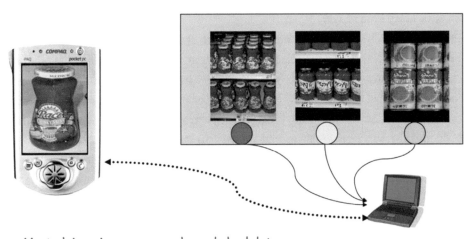

Matching images on hand-held to screen

Figure 3.6: Image /icon recognition experiment setup. From Carmien and Wohldmann (2008).

Participants were tested individually in a quiet room. Each trial consisted of matching a picture displayed on the hand-held computer to one of three pictures displayed on the white screen by pressing one of three buttons that were placed in front of each picture on the white screen. Response time was measured as the time that participants took to press one of the three buttons after the onset of the image on the white screen.

Each image (called the cue) was scaled to fit on the hand-held computer (280 × 210 pixels), and the images displayed on the white screen included the correct target, a target lure (an item that could possibly make for a suitable substitute in place of the target), and non-target item (an item that was neither the target nor a near substitute for the target, but had a similar shape or size to the target) (Figure 3.6). For example, if the target image was jar of spaghetti sauce, then the target lure might be a jar of salsa and the non-target might be a jar of peanut butter. Each target could be represented on the hand-held computer in one of three ways. Specifically, the picture was either shown in isolation (a single target object shot against a white background), in context (a view of the target item on a grocery shelf), or was represented as an icon from the set of Mayer-Johnson picture-symbol icons (Mayer-Johnson, 2001) that are typically used in instruction for this specific population of young adults.

A short break was given after every 10 trials were completed, and for each participant the experiment consisted of 2 blocks of 30 trials with 10 of each picture type (icon, isolated, context) in a block. Each block had 10 sets of targets paired with one of the three kinds of cue images on the PDA. The 30 trials were randomly scattered across a block. In this way every participant was presented with every kind of cue for a given target.

The experiment began with a demonstration to show how the task was to be performed. After this initial practice phase the hand-held computer was set up for the trial. The hand-held computer had two states—currently showing the target image and ready to show the next target. The participant activated these states by touching the screen. Once the set of pictures had been selected, the hand-held showed the "ready" image. The participant was then asked to touch the screen when ready and the hand-held transmitted a signal to the laptop, triggering the laptop to both import the images that the participant would see on the projector screen and to send a signal back to the hand-held to display the target image on its screen. The images appeared simultaneously on the hand-held computer screen and on the white screen. When the participant pressed one of the three buttons, the computer obtained information identifying the image and a code that indicated whether that image was a perfect match, a lure match, or a non-target match.

Two dependent measures were included in the analyzes. First, response time was calculated as the time that it took participants to press one of the three buttons after the pictures were displayed on the white screen. Only accurate trials were included in the response time analyzes. Second, accuracy of the response was calculated and reported as proportion correct. The overall design for this experiment was a mixed factorial including participant type (cognitively disabled, cognitively typical) as a between-subjects variable and block (first, second) and target type on the hand-held computer (isolation, context, icon) as within-subjects variables.

A repeated measure ANOVA was conducted on both response time and accuracy. Two types of analyses were conducted on each measure. The first type included all three target types (isolation, context, icon) for comparison. The second type compared icons to the average of realistic images (i.e., to the average of those in isolation and in context) to determine how suitable icons are for representing real-world items. One typical subject was removed from the analysis because his accuracy performance was more than four standard deviations away from the mean of his group.

Both the first and second analyses of response time yielded a significant main effect of participant type. Typical participants showed faster response times compared to cognitively disabled participants. No effects involving picture type were significant.

In both types of analyses conducted, partially correct (i.e., the selection of a target lure) trials were included because target lures would make a suitable substitute for the target item and because there were very few correct trials from the participants with cognitive disabilities. Thus, correct trials were given a score of 1, correct target-lures were scored as .5, and non-targets were scored as 0.

As expected, the first type of analysis yielded a significant main effect of participant type. Specifically, typical participants were more accurate on the matching task compared to those with cognitive disabilities. In addition, participants were more accurate when the object was represented realistically either in context or in isolation than when it was represented as an icon; the main effect of picture type was significant.

In the second analysis comparing icons to realistic images, the main effect of participant type was also significant. Again, typical participants were more accurate than those with cognitive disabilities. In addition, the main effect of picture type (icon, realistic) was significant, with, overall, higher accuracy on realistic images compared to icons. Furthermore, the interaction of participant type and picture type was marginally significant. Participants with cognitive disabilities were less accurate when the pictures were represented as icons than when they were represented realistically, and although typical participants showed a similar trend, the difference in accuracy between icons and realistic images was smaller. This finding suggests that icons may be poor substitutes for realistic representations, especially for people with cognitive disabilities.

There was no significant difference between the two types of realistic images (isolated and in context). This was unexpected, as existing research (Braun, 2003) tends broadly to emphasize context and contextual cues as significant in object recognition. There are at least two possible explanations for our findings. First, the images called "in context" might not be in context in the way that the other studies used the word. Second, in the case of selecting discrete objects in the world on the basis of a small two-dimensional image, context may not be very pertinent.

Regarding the second explanation, it might be that what many of the other studies were looking for was the "meaning" of the images, especially with respect to communicative intent (for

AAC purposes). For this specific use of images, and especially in a use environment where the image used is intended to be very specific to the goal, such as in a supermarket (Carmien, 2006b), the fidelity of the image is much more important than its arrangement. In this case of the MAPS system it is easy to change the images to reflect the specifics of the task on a day-to-day basis (i.e., the image could be changed for each instance of the prompter script that fit the user, goal and task). In any case, our results, while not conclusively supporting the conjecture that drove the experiment, suggest that further exploration of this topic would be helpful in building an effective computational task support system.

Based on the results, icons and photographic images depict real-world objects in different ways. This difference, whatever it may be, is enough to lead to errors in matching those representations to their real-world referent. The results imply that making "generic" scripts out of icons would not be a good strategy for creating scripts to guide a person with cognitive disabilities in daily tasks. Given that knowledge, the current practice of assembling prompts out of icons needs to change toward using realistic photos. For some participants and objects a generic photo may suffice, for others a generic photo may either not be available or not have rich enough detail to afford easy matching. This leads to a requirement of computationally enhanced prompting systems to support the easy use of caregiver-taken photos in the creation of scripts.

3.2.3   CONCLUSION

The importance of representation on interfaces is critical to success in applications designed for AT/IA. This, as discussed above, is an expression of the cognitive abilities of the target end-users. Also, and perhaps not so obviously, is the choice of navigation and controls for the functions of the application. Often, novice designers will try to express their creativity by designing new icons for actions like getting help or starring a action, this can lead to failure in two ways: (1) a programmer is not a trained graphic designer and the results, while pleasing, may be incomprehensible to the end-user; and (2) introducing novel icons to replace ones that the end users, especially young adults, may already understand, creates another cognitive hurdle to surmount. Reuse successful interface elements and think (and do trials with end-users) before injecting improvements. In fact, this is a useful guide line for all user interface design (Lewis and Rieman, 1993).

3.3   TOOLS FOR LIVING AND TOOLS FOR LEARNING

Short Definition: Tools for learning are systems that support changing the user to (re)gain skills. Tools for living are systems and devices that support the end user in doing tasks that they cannot do. Tools for learning are used and (hopefully) abandoned; tools for living are carefully fitted to the user and typically used for the rest to the user's life (Carmien and Fischer, 2005).

Longer description: Tools for *learning* support people in (re)learning a skill or strategy with the objective that they will eventually become independent of the tool. Tools for *learning* afford an internalization of what was (if it existed previously at all) an external ability/function; tools for learning often serve a scaffolding function (Pea, 2004).

In contrast, tools for *living* are external artifacts that empower human beings to do things that they could not do by themselves. One use of a tool for living is supporting distributed cognition; that is, it serves as an artifact that augments a person's capability within a specific task without requiring the individual to internalize the sub-tasks conducted by the artifact (e.g., a hand calculator). Tools for living can be tailored for specific tasks and for specific people.

CLever's ("Cognitive Levers: Helping People Help Themselves") (CLever, 2004) goal was to create more powerful media, technologies, and supportive communities to support new levels of distributed cognition (Carmien et al., 2005). This support is designed to allow people with disabilities to perform tasks that they would not be able to accomplish unaided. The objective is to make people more *independent* by assisting them to live by themselves, use transportation systems, interact with others, and consistently perform normal domestic tasks such as cooking, cleaning, and taking medication.

CLever identified and explored a fundamental distinction in thinking about the empowerment and augmentation of human beings (Engelbart, 1995) and the change of tasks in a tool-rich world by identifying two major design perspectives.

1. *Tools for Living:* grounded in a "distributed cognition" perspective, in which cognition is mediated by tools for achieving activities that would be error prone, challenging, or impossible to achieve.

2. *Tools for Learning:* grounded in a "scaffolding with fading" perspective leading to autonomous performance by people without tools.

Internal and External Scripts: The differentiation between tools for living and tools for learning is closely related to a similar fundamental issue: the interplay and integration between *internal and external scripts* (Carmien et al., 2007). *Internal scripts* are chunks of mastered acquired behavior that can be executed without the need for external support. *External scripts* are instructions that afford the accomplishment of more complex tasks by triggering internal scripts to execute the externally cued steps. Tools for living exploit the *interplay and integration between internal and external scripts*. In contrast, the goal of tools for learning is to acquire new internal scripts, thereby becoming independent of external scripts.

3.3.1   TOOLS FOR LEARNING

Looking at the relationships between humans and the artifacts they use, tools or systems can be seen as primarily either supporting human adaptation or providing support for humans to affect change in their environments. Thinking about systems and artifacts in this way affords insights into distributed intelligence, the design and use of artifacts, and educational decisions about learning and teaching of skills with and without tool support. Some artifacts support people by providing guidance and then gradually removing assistance until the skills become internal scripts, accessible without external assistance. Examples of such tools for learning are: (1) bicycles with training wheels (see Figure 3.7); (2) toddlers' walkers; (3) wizards (environments that interactively guide users through the steps that comprise the process of setting up computer applications) used in many computational environments; (4) experienced teachers providing help and assistance to learners; and (5) caregivers assisting people with disabilities.

Figure 3.7: Learning to ride a bicycle with training wheels. An example of a tool for learning. Courtesy of s_oleg/Shutterstock.com.

Optimally, a tool for learning uses scaffolding of some sort to provide enough support in the beginning of the tool use to actually compete a task without so much frustration that the user quits, an example of which are systems to teach programing (Wood et al., 1976) or the scientific approach (Guzdial, 1994). Similarly, it should have some sort of graduation point where the tool is no longer needed to do the task, an example of this is the removal of the training wheels on a bicycle.

Tools for Living

Some artifacts enable users to perform tasks that are impossible for them to do unaided. Often, no matter how many times the task is accomplished with the aid of a tool, the user has no greater ability to do the task unaided than she or he did initially. Examples of these tools for living include ladders, eyeglasses, the telephone, screen readers for blind people, visualization tools, and adult tricycles (see Figure 3.8). No matter how many times people use the phone to talk to friends across town, their native ability to converse over long distances unaided remains the same as before they used the telephone. Tools for living allow people with disabilities to perform tasks that they would not able to accomplish unaided, and therefore allows these people to live more independently.

Figure 3.8: Adult tricycle: an example of a tool for living. Courtesy of Belize Bicycle, Inc.

Tools for living require a tight fit between the user and the device (e.g., eyeglasses and eye examination). They also are used forever (exceptions being laser surgery and progressive loss of vision).

Trade-Off Analysis: As with any category scheme, there are overlaps and differences of interest and use (Table 3.1 summarizes some of those). The categorization of devices depends only, to a small extent, on the intentions and viewpoints of designers and developers, but their use is much more profoundly determined by the cultural values of the use situations and the societal organizations that determine them. The inappropriate use of some tools for living has been identified as causing a sort of "learned helplessness" in that the ease and accessibility of using some of these tools inclines the user to not expend the energy and time to acquire these skills internally. Examples of this learned helplessness are the use of hand calculators and spell checkers as tools for living, thus blocking the acquisition by the users of arithmetic and spelling skills, not unlike the above-mentioned example of complaints about reading ruining native human abilities. Much has been written and debated about the use of calculators, spell checkers, and other cognitive tools in education and whether they should be used as tools for living or tools for learning and how these decisions may be different in the case of children who suffer from permanent disabilities such as dyslexia.

Table 3.1: Overview of tools for living and tools for learning

	Tools for Living	Tools for Learning
Definition	Use to do task with tools	Learn to do tasks without tools
Examples	Eyeglasses, phone, radar, cockpits, scuba diving gear	Spelling correctors, hand-held calculators, tricycles, wizards
Strengths	Extends reach beyond natural ability	Extends unaided reach
Weaknesses	Learned helplessness	Long training time (reading can take thousands of hours to learn)
People with Disabilities	Optical character recognition page readers, specialized mice and keyboards, assistive communication devices, adult tricycles	Learning Braille, learning how to use prompts and prompter, learning a specific bus route
Prompting Systems (maps)	For people with memory problems (disabilities, elderly)	For people without memory problems (but: people use calendars on paper, reminding systems)
Distributed Cognition Perspective	Resource rich (professional life)	Independent of external resources (school), memorization

Over-reliance on Tools for Living: Does an over-reliance on tools for living lead to learned helplessness and deskilling, ruining the user's native abilities by making them dependent on the tool? The cautionary point of *The Time Machine* (Wells, 1905) was that the Eloi had lost any power in their world by having too much done for them. It is this fear that leads to a moral judgment about the use of such tools. Asimov (1959) tells the thought-provoking story "The Feeling of Power," in which a scientist who is actually capable of doing simple math problems on paper without the aid of a computer is identified by the military of that distant century and how the military tries to make use of his "newfound skills." Perhaps the resistance to them is also tied to the 19th century belief that mental deficits were illnesses that reflected moral vices or immoral lifestyles. The logic goes something like this: a calculator is a "crutch" that allows a person to forgo the effort associated with true learning (displaying the moral vice of laziness). There is little doubt that the inappropriate use of these devices can lead to "learned helplessness" (e.g., providing screen readers to children with dyslexia has in some cases prevented them learning techniques that may have largely ameliorated the effects of their developmental disability (Olsen, 2000)). However, accepting that as a genuine concern does not necessarily lead us to the conclusion that all such technology is bad for humans. The daily life of all of us is filled with an almost infinite number of tools for living.

3.3.2   CANONICAL PAPER

Carmien, S. and G. Fischer (2005). Tools for living and tools for learning, (Carmien and Fischer, 2005).

3.3.3   AT EXAMPLES

An other example of a tool for living, but having some attributes of a tool for learning and more situated in AT for IA, was the MAPS project. This is a good example because of the parts that were designed from each tool's perspective highlights the tool's approach well.

Prompting has long been used as a technique to aid people with cognitive disabilities to live more independently. With the advent of computationally based prompting systems, the image and verbal prompts that lead a user through task completion can now be highly personalized, and the need to memorize the steps to a task is off-loaded to the computerized prompter (in this case, an HP IPAQ h6315), thus changing the task from memorizing scripts of prompts to accessing the appropriate script and following the prompts as they are presented. Our prompting system consists of the caregiver's script editor, which supports the creation and storage of scripts consisting of pairs of images and verbal prompts, and a handheld prompter, which is used by the person with cognitive disabilities and presents the previously made scripts.

MAPS as a Tool for Living

MAPS was designed as a tool for living inspired by the Visions system (Baesman, 2000), a stationary prompting and scheduling system based on PCs using speakers and stationary touch screens to prompt a user with cognitive disabilities through a complex domestic task such as cooking. Part of the Visions system supplied sets of cards that assisted away-from-the-system tasks such as grocery shopping. Although Visions worked, it had problems with re-configuration, and the computationally implemented in-home video/audio prompting component was not able to go with the user out of the home in which it was installed (other than the sets of cards). As a result of the problems with re-configuration and generalizability issues, the MAPS user interface was made for two (end) users: the actual user and the caregiver. As a tools-for-living artifact, the initial setup was critical, and special care was paid to functions that aided the caregiver in conforming to the needs and abilities of the prompter user. The initial configuration supports specifying image and verbal prompt styles: image aspect, verbal prompt complexity, and word choice. The configuration wizard also parameterizes the capturing and correction of errors: how to detect that a script was not being followed, how to detect the problem that caused the error, and what to do to correct the error (some users can assist with self-correction, some will need human intervention (e.g., activating the MAPS cell phone connected to a caregiver, and some will require emergency intervention (e.g., calling 911)).

Because the target population has a limited number of possible "operations," the set of available scripts should not vary greatly, and the same scripts will be used over and over. Moreover, many scripts for outings will be constructed by reusing sub-scripts (e.g., the steps to "get from getting ready to going out to the closest bus stop" will be repeatedly used). What will change is the timing and, to a small degree, the content of the scripts. For example, if MAPS were equipped with a GPS and is networked so that when the user gets to the bus stop, the arrival of a specific bus will trigger the prompt to get on the bus (Sullivan, 2005) (see references to the ASSISTANT system which implemented this). As a tool for living, MAPS is expected to support a specific, small set of tasks and, in this mode, not provide support for autonomous generalization of the skills through which the user is being guided.

Training in using MAPS, just as training in using any tool for living, consists of a simple set of instructions in the use of the tool, not extended to any overview or understanding in the principles involved in doing the prompted through task itself. Similar to instructions for the care and use of contact lenses, in which there is no attempt to explain optics, training in using MAPS does not involve anything about the domain of the tool other than pragmatic ATM-style instructions. This is typical for the change in the task that using tools for living involves. The task now is the use of the tool, not the difficult cognitive tasks of figuring out what to do and when to do it. This task re-mapping is a typical result of bringing a tool for living into play. Another example of this is the change in the details for arithmetic calculator users, who are now experts on using the buttons to input and control the calculator rather than memorizing complicated algorithms for calculating square roots. In this aspect the AT/ IA versions of Tools for Living implement many of the concepts of distributed cognition (see Section 2.3).

Concerns about the robustness of the tool, and concerns about the target population (e.g., what if the user loses it, what if it breaks while the user is in a bad part of town at night?) are part of our research agenda. Any design that allows and encourages a somewhat fragile population into a situation that may be somewhat of a threat in the case of breakdown has a moral responsibility to ensure an elegant decay/failure in such cases. LifeLine (Gorman, 2005) aims to address these concerns by monitoring the use of MAPS and involving a caregiver if necessary (Carmien et al., In Press). These issues were also addressed in the ASSISTANT system (see Section 2.2).

Creating Tools for Living: A crucial part of MAPS is the script design environment used by caregivers to design effective external scripts that trigger the internal scripts to allow people with cognitive disabilities to live more independently. Critical to successful design of a tool for living is the initial fit of the device; examples include visits to an optometrist for eyeglass fitting or to a clinic for prosthesis fitting—devices that require precise alignment to the user to operate. In the case of the MAPS system, the hand-held user needs images that have the right affordances (Norman, 1993) to allow the correct selection of the desired object. Similarly, the verbal prompts need to have the right level of words and grammar to fit the needs and abilities of the person with cognitive disabilities that is using the prompter. Generic icons and computer-generated verbal prompts will typically not be as effective as ones tailored to the user specifically (Snell, 1987). Figure 3.9 shows the caregiver's script editor in use: the caregiver builds prompts by selecting from sets of pictures and recorded prompts, and creates scripts from sequential prompts The finished script can then be loaded into the prompter for use by the person with cognitive disabilities in accomplishing the desired task. Because this initial fit is so critical to the use of the tool for living, generic script templates, which have been previously used successfully, are provided for caregivers to use as models in producing specific scripts.

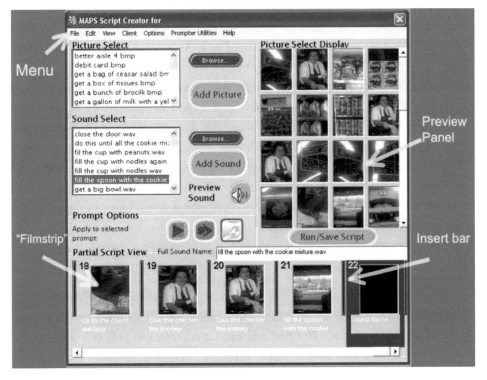

Figure 3.9: MAPS script editor sections. From Carmien (2006b).

Learning to Use Tools for Living: Tools for living will be of no help if people are unable to use them. In the context of MAPS, the system explored the interpretation of specific external scripts. In designing and using tools for living, two key elements of which are the initial fit and the correct leverage of existing skills. In the case of MAPS, the caregiver and assistive technologist need to understand what would be the best modality and style for the effective delivery of scripts. Perhaps the user is able to map standard icons to articles in the real world for a shopping script; this would make the creation of scripts much easier than those for the person with cognitive disabilities who needs specific pictures of the exact target item. Similarly, some users would respond best to prompts from a familiar authority figure (i.e., mom), whereas for others this might be totally wrong, due to typical teenage power struggles (the user's own voice doing the prompting might be best). Having determined the appropriate modality for the tool, the caregiver needs to assess and fit the prompting structure of the script for the internal scripts that the user may have. For instance, one user may need only an overarching prompt of "*Go to your bus stop*" whereas another may need to have this segmented into three steps: "*Walk down the street to the corner,*" "*Cross the street,*" and "*Turn left and walk to the bus stop in the middle of the block,*" each one of these steps corresponding to an internal script possessed by the user that may be successfully triggered by an external script in the prompting sequence.

Another perspective on using tools for living is that once the task that was too difficult to do is remapped to a task that is within the domain of existing skills, there still remains the need to acquire those skills that are needed for the remapped task. An example in the case of MAPS would be replacing the whole task of performing a janitorial job at a restaurant, requiring a series of executive (when to do the subtask) and mnemonic (the specifics of the subtask) actions that constitute the job, with the task of using the controls on the prompter to sequentially trigger the existing internal scripts that would then constitute the whole job. An example from another domain might be replacing the task of memorizing the Qur'an with reading the Qur'an. In both examples there is the need to learn the use of the tool for living, and this learning with the same external support constitutes a tool for learning. In the case of learning the use of the MAPS prompter, studies have observed young adults with cognitive disabilities taking about a half hour; in the case of learning reading, this would take perhaps many thousands of hours.

MAPS as a Tools for Learning

MAPS is also designed to be used as a tool for learning; even when used in the tools-for-living mode, it will afford some tools for learning functionality. MAPS is designed to accommodate several types of scaffolding fading. The first and most crude form of fading is based on a review of the logs of user behavior that MAPS keeps (with awareness of the privacy issues involved): the caregiver can consequently manually edit a given script, folding several prompting steps into one "reminder" prompt. Next, MAPS could evaluate usage and autonomously, in an algorithmic fashion, compress a script based on logs in a batch mode. Finally, MAPS could dynamically expand or contract the steps required to perform a given task based on past behavior, user input, and immediate history. Neither of these were actually implemented in the project, but were simulated by hand and an algorithm generated with an aim of later implementation. MAPS is designed to provide scaffolding to augment memory and cognition; as a memory prompt tool, it may just need to "nudge" the user in the correct direction upon determining that an error condition is about to happen (a probabilistic guess, but no harm done if the guess is wrong), Suchman (2007) discussed this in the second edition of her seminal work, *Plans and Situated Actions*.

3.3.4   CONCLUSION

Problems in using the wrong tool include, in an example using a tool for learning in place of a tool for living, spending inappropriate effort in trying to learn the task instead of using a tool for living to just make progress. This trade-off is based on the amount of time/resources needed to learn the task (e.g., using a calculator versus learning numerical analysis in a computer science engineering school to do cube roots). Similarly using a tool of living in the place where a tool for learning is more appropriate can make a task frustratingly slow and motivate abandonment. An example of this misapplication of class of tool is giving a system to compensate for dyslexia to a younger person where it might be more appropriate to provide a set of exercises so that they can internally compensate for their reading problems, one that they can scaffold and graduate from (Olsen, 2000).

Problems in using the tool wrong include deskilling inappropriately, an example of which is losing the ability of doing simple arithmetic in your head as a result of using a calculator exclusively. (A problem I had after undergraduate engineering school, which I worked successfully on by not bringing out my calculator or pencil for simple problems.) Another example is losing a mental overview map of a route by overreliance on GPS.

The most easily recognized tools for living—eyeglasses and prosthetic limbs—are clear examples of the properties of these tools. Supporting tasks that will never be doable without them, precise fitting, and never graduated from (i.e., used forever).

There are, as mentioned above, many stratagems and tools for learning compensatory skills for moderate dyslexia (Olsen, 2000). Similarly, rehabilitative systems for re-mapping muscle functions in stroke victims are in contrast to AT provided later when the brain is not so amenable to exploiting its plasticity (Perry et al., 2009).

CHAPTER 4

Technique

This chapter describes a set of approaches that help make an AT/IA system work with a tractable design to implement.

- Plans and actions

- Low-hanging fruit

- MetaDesign

- Personalization

- Symmetry of ignorance

4.1 PLANS AND ACTIONS

Short Definition: A script is a description of how a goal is attained, and scripts can be broken down into sub-scripts and eventually into "atomic" steps. Attempting to follow a script can be unsuccessful in many ways, which are called errors.

Longer Description: Early in the development of cognitive science's discussion of human behavior in describing day-to-day behavior, a model emerged of scripts, goals, and actions. Basically this breaks down a set of behaviors into steps that, combined, are called scripts. Scripts are aimed at a goal, and can be broken down into sets of sub-scripts with intermediate goals until "atomic" actions are reached. These ideas came out of both studies in education (Weinberger et al., 2005) and early AI and cognitive psychology (Schank and Abelson, 1975). First some definitions:

> **Goals** are the desired end point of the script. Important to defining and discussing goals is that they be measurable, that there is an endpoint. Also important, especially from an AI perspective is the use of goals as part of a utility function in deciding the best action (Russell and Norvig, 2009).

> **Plans** are the description of how to get from the current state to the goal state. As with goals, AI literature has much to say about this topic. One of the critical issues with plans is how they are represented. The plan could be an algorithm, a set of cards, or computer-supported prompts. Plans are typically comprised of steps of appropriate granularity.

Actions are the constituent part of plans that lead to the goal. Actions can be sub plans, can be cued externally, and can be an inventory of internal abilities.

Errors are typically included in the canonical literature is the notion of a breakdown in the process. This can be referred to as an error. Other sections of this book discuss this in some detail (see Section 2.2).

Schank and Abelson (1977) described human behavior in terms of goals and scripts. A script is a description of how a goal is attained. These scripts are typically internal to the person acting on them. However they can be looked at another way. In training and rehabilitation of a person with cognitive disabilities, professionals often train people with cognitive disabilities to accomplish ADLs by splitting ADL tasks into discrete steps that the client can easily accomplish and then teaching the task as a set of these kinds of tasks sequentially linked together (Saskatchewan Learning—Special Education Unit, 2003). The client then memorizes this sequence of steps and then has the ability to do a task that she could not do previous to the training. Such a set of steps is called a script and the individual steps are called prompts. By moving this process out of internal representation into a computationally based system the learning is moved from that to learning the task scripts one by one to learning how to use the script player (implemented on a PDA); such a transfer of necessary skills is part of the process of distributing of cognition (Solomon, 1993; Hollan et al., 2001). Distributed cognition is also in play in the change from memorizing long texts (i.e., the *Qur'an*, The *Iliad*) to acquiring the ability to read these texts in written form; thus acquiring the distributed cognition skill of reading effectively enables a person to "memorize" as many texts as there are books. In this case the steps are presented, as prompts constituted from an image and a verbal instruction. Tailored exactly to each person with cognitive disabilities, some tasks are represented with many more prompts (corresponding to the clients lack of internal sub-task knowledge) or less prompting steps (corresponding to a client having more internal sub tasks) (Carmien, 2006b). So the client's limited internal scripts are complemented by powerful external scripts (Carmien, 2006b).

One advantage to looking at task support in this way is that it becomes easier to create guides and tutorials for thecreation of scripts using task segmentation. Task segmentation is the process, taken from the domain of vocational rehabilitation, of breaking a specific task, with granularity of prompts derived from intimate knowledge of the intended end-user's needs and abilities, into a set of steps, each of which can be done as an "atomic" task. Here, atomic means that the sub task is itself an internal script that the end-user can do without support. For some this is larger, like get the ingredients for the sandwich from the refrigerator, for others smaller, like go to the refrigerator, open the door, remove the mayonnaise.

The practical implementation of the notion of plans using scripts for defining a task and supporting the successful accomplishment of the task comes to this domain from the world of rehabilitation. There is a plethora of practical best practices and already prepared templates for tasks for people with cognitive disabilities. These templates become very useful for showing how to segment a task as well as a stepping off point for creating individualized versions. The LRE for LIFE Project (LRE for LIFE Project, 2001) provided a set of scripts to be used by professionals in teaching activities to people who cannot currently do them, ranging from simple to more complex (see Figures 4.1 and 4.2)

Name: _____

Goal: __Undressing_____   **Data**:   / = correct response
_____   X = incorrect response
                                                                          P = prompted response
                                                                          Circle Total Number Steps Correct

Natural Cue(s): _____   **Natural Consequences**: _____

step	Step/Behavior						
7	Put shoes up	7 7 7 7 7	7 7 7 7 7	7 7 7 7 7	7 7 7 7 7	7 7 7 7 7	7 7 7 7 7
6	Replace garments in appropriate place	6 6 6 6 6	6 6 6 6 6	6 6 6 6 6	6 6 6 6 6	6 6 6 6 6	6 6 6 6 6
5	Put on nightwear or other appropriate clothing	5 5 5 5 5	5 5 5 5 5	5 5 5 5 5	5 5 5 5 5	5 5 5 5 5	5 5 5 5 5
4	Remove undergarments	4 4 4 4 4	4 4 4 4 4	4 4 4 4 4	4 4 4 4 4	4 4 4 4 4	4 4 4 4 4
3	Remove outer garments	3 3 3 3 3	3 3 3 3 3	3 3 3 3 3	3 3 3 3 3	3 3 3 3 3	3 3 3 3 3
2	Remove shoes	2 2 2 2 2	2 2 2 2 2	2 2 2 2 2	2 2 2 2 2	2 2 2 2 2	2 2 2 2 2
1	Go to appropriate area	1 1 1 1 1	1 1 1 1 1	1 1 1 1 1	1 1 1 1 1	1 1 1 1 1	1 1 1 1 1
		0 0 0 0 0	0 0 0 0 0	0 0 0 0 0	0 0 0 0 0	0 0 0 0 0	0 0 0 0 0
	Codes: Location						
	Date						

Figure 4.1: Script to teach undressing. From LRE for LIFE Project (2001).

Name: _____

Goal: _Making breakfast foods/waffles_   Data: / = correct response
_____   X = incorrect response
                                                                                                                 P = prompted response
                                                                                                                 Circle Total Number Steps Correct

Natural Cue(s): _____   Natural Consequences: _____

step	Step/Behavior						
16	Dishes away	16 16 16 16 16	16 16 16 16 16	16 16 16 16 16	16 16 16 16 16	16 16 16 16 16	16 16 16 16 16
15	Clear table/wash dishes	15 15 15 15 15	15 15 15 15 15	15 15 15 15 15	15 15 15 15 15	15 15 15 15 15	15 15 15 15 15
14	Unplug toaster	14 14 14 14 14	14 14 14 14 14	14 14 14 14 14	14 14 14 14 14	14 14 14 14 14	14 14 14 14 14
13	Syrup away	13 13 13 13 13	13 13 13 13 13	13 13 13 13 13	13 13 13 13 13	13 13 13 13 13	13 13 13 13 13
12	Eat	12 12 12 12 12	12 12 12 12 12	12 12 12 12 12	12 12 12 12 12	12 12 12 12 12	12 12 12 12 12
11	Syrup on waffles	11 11 11 11 11	11 11 11 11 11	11 11 11 11 11	11 11 11 11 11	11 11 11 11 11	11 11 11 11 11
10	Plates on table	10 10 10 10 10	10 10 10 10 10	10 10 10 10 10	10 10 10 10 10	10 10 10 10 10	10 10 10 10 10
9	Waffles on plates	9 9 9 9 9	9 9 9 9 9	9 9 9 9 9	9 9 9 9 9	9 9 9 9 9	9 9 9 9 9
8	Set table (forks, napkins syrup)	8 8 8 8 8	8 8 8 8 8	8 8 8 8 8	8 8 8 8 8	8 8 8 8 8	8 8 8 8 8
7	Get plates/put by toaster	7 7 7 7 7	7 7 7 7 7	7 7 7 7 7	7 7 7 7 7	7 7 7 7 7	7 7 7 7 7
6	Put waffle box away	6 6 6 6 6	6 6 6 6 6	6 6 6 6 6	6 6 6 6 6	6 6 6 6 6	6 6 6 6 6
5	Push lever down	5 5 5 5 5	5 5 5 5 5	5 5 5 5 5	5 5 5 5 5	5 5 5 5 5	5 5 5 5 5
4	Put in toaster	4 4 4 4 4	4 4 4 4 4	4 4 4 4 4	4 4 4 4 4	4 4 4 4 4	4 4 4 4 4
3	Get waffles/open box	3 3 3 3 3	3 3 3 3 3	3 3 3 3 3	3 3 3 3 3	3 3 3 3 3	3 3 3 3 3
2	Plug in toaster	2 2 2 2 2	2 2 2 2 2	2 2 2 2 2	2 2 2 2 2	2 2 2 2 2	2 2 2 2 2
1	Wash hands	1 1 1 1 1	1 1 1 1 1	1 1 1 1 1	1 1 1 1 1	1 1 1 1 1	1 1 1 1 1
		0 0 0 0 0	0 0 0 0 0	0 0 0 0 0	0 0 0 0 0	0 0 0 0 0	0 0 0 0 0
	Codes: Location							
	Date							

Figure 4.2: Script to teach cooking breakfast. From LRE for LIFE Project (2001).

With these as a guide and some simple short exercises the caregivers in the MAPS project became proficient in segmenting tasks for their end-user participant.

In many practical AI problems, a hard problem can be made tractable by constraining some of the parameters of the problem. Even after making the problem less "wild" this can still be a daunting task. As an example, the solution of a constrained problem, supporting the washing of hands in an instrumented bathroom by Alzheimer's patients with error correction and plan identification, took five years and many researchers from different disciplines

There are two interesting problems associated with task support. The first one is to infer a plan and its goal (or a goal and its plan) on the basis of perceived actions of the end user. This problem is a classic AI issue and has been extraordinarily difficult to solve in a general sense. When this topic was discussed in L3D, one of the guest scholars commented on the difficulty of inferring a goal from a set of observed actions saying many times he himself did not really know what he was trying to do. This is also a common dead-end that computer scientists make when approaching dynamic and complex task support, starting with the assumption that it was possible to dynamically ascertain what the user was trying to do by observation and sensor inputs. This is a hard problem and, while a worthy goal, is really not an AT/AI project that will succeed in the near future. As discussed above, one way to make such a problem tractable is to constrain the problem space, which was what allowed the COACH project to succeed (i.e., in an instrumented bathroom, one simple task).

4.1.1 CANONICAL PAPER

Schank, R.C. and R.P. Abelson (1977). *Scripts, Plans, Goals, and Understanding.* (Schank and Abelson, 1977).

4.1.2 AT DESIGN EXAMPLES

Script Segments and Plan in MAPS

To use MAPS support for performing a task, first someone needs to make a set of prompts, but to do this several things need to be prepared: (1) the elements of the task need to be separated into individual sub-tasks, and (2) these subtasks need to be based on the existing abilities of the end-users. In some cases the subtasks will be large (e.g., a script to make a sandwich consists of three prompts—get the bread and sandwich ingredients from the refrigerator, make the sandwich, clean up); in some cases the prompts may be many more (e.g., go to the refrigerator, get the bread, put the bread on the countertop, go back to the refrigerator and get the package of ham and the mayonnaise, put them on the counter ... and on). All of this is dependant on the available set of "internal" scripts that the end-user has available. In this case the element of intelligence augmentation is the fit to the user's support needs, where simple AI might simply provide the script that breaks the talks into as many steps as possible, one size fits all, the IA approach tailors the script to fit the user. Table 4.1 is a segmented task that was created to evaluate the usability of the hand-held prompter.

	Prompt	Image
	Table 4.1: MAPS testing and training script	
1	Take the glider parts out of the bag.	Clear plastic bag with parts
2	Make sure you have all five parts.	Parts you laid on table top next to bag
3	Slide the big wing through the big slot in the plane until it is in the center. Be careful when you slide it in.	Just this w/ hand
4	Slide the small wing through the small slot in the back of the plane.	Just this w/ hand
5	Put the tail wing on top of the back of the wing in the small slot.	Just this w/ hand
6	Put the pilot on the top of the plane.	Just this w/ hand
7	Hold the plane with your fingers and throw it.	Just this w/ hand
8	Have fun, the glider is yours to keep.	Glider in mid-air.

The COACH project, as discussed in Section 2.2, particularly in Figure 2.15, used the plans and actions model to base its handwashing support system.

4.1.3   CONCLUSION

Plans, scripts, and actions are all related to discrete tasks. When designing supports for accomplishing a specific and discrete task, the process of segmenting the task into steps can be useful, even if you are not specifically making a system similar to the above task support. For instance, in the ASSISTANT project, informally looking at the elder's route guidance needs in this way lead the design group to a useful set of modalities that, put into a petri net-like structure (e.g., a finite-state machine), helped a lot in working our the Design for Failure systems in the project.

4.2   LOW-HANGING FRUIT

Short Definition:

> *"Le meglio è l'inimico del bene"*

> *"The best is the enemy of the good"* (Voltaire; Peter Gay (1962)).

Approaching complex systems, by attempting to solve the whole problem, a designer may end up solving no problem. Complex AT should be carefully evaluated with respect to the actual needs of the user and available resources of the situation and often it is more successful if the easier and more common parts are addressed initially.

Longer Description: The standard way of developing a system starts with an idea and proceeds with requirements gathering. Based on interviews and focus groups with end-users, experts, and other stakeholders, technologists attempt to come up with a design that will surpass the effectivity of existing work practices. This is a good thing, however in our training as professionals we had drilled into us the need to conceder edge situations in design and implementation. A rough way to evaluate code as ready for the real world is counting cases of "if-then" statements that do not have other cases. We think robust; we seek coverage of the problem in beautifully complex and complete systems. However, this may work against us in real complex AT projects, as the solution may be too complex to use or may depend on somewhat fragile infrastructure that is efficacious in the lab and not so much in the "real world." Attempting to do it all can often in complex applications lead to failure of the whole system. Furthermore, most projects of this kind are wicked in nature (see Section 2.8), with the consequence that the real requirements may not come to light until an initial prototype deployment. From this insight, we often referred in the L^{3}D lab as "low-hanging fruit" as an initial goal, trying to keep the complete solution in mind so as not to prevent, by the architecture, later passes at the system that might implement more covering solution.

So, how to prevent the perfect preventing completion of good in high-functioning AT? One of the tools I have brought to this domain from a previous career as a material control manager in the 1970's and 1980's was the idea of controlling inventory with what is called in production and inventory control management as "ABC" analysis. At the time I was working in this discipline we were trying to emulate the Japanese, who had become experts in inventory control and just-in-time manufacturing following the approach of W. Edwards Deming (1950), whose ideas were rejected in the U.S., leading him to present them to the Japanese manufacturers. Coming out of his statistical approachs were sets of tools that supported efficient work practices to raise inventory accuracy in factories from typically 75% to the low 90% range. One of the tools that came out of the (eventually successful) attempts at total quality control and just-in-time manufacturing management, was striving to get very high accuracy of the raw and intermediate materials that comprised a cumulatively high cost in the finished product. Attempting to do this with all the parts and assemblies in the factory at once was too expensive to implement for all parts as a result of diminishing returns (i.e., there were items that were so relatively inexpensive that driving the records to 95% accuracy was more expensive in manpower than just buying more than you think you need). You don't count individual paperclips—you make sure you always have more than enough—you do count automobile engines.

Based on Pareto's[11] observations that, in many fields of human endeavour, 80% of the value of the items being inspected is accounted for by 20% of the item types. It's easier to see this in the

---

[11]   Vilfredo Federico Damaso Pareto (1848 –1923), Italian engineer, sociologist, economist, political scientist, and philosopher. He made several important contributions to economics, most meaningful in this book was his study of income distribution.

context of the current concern about almost half of the world's wealth is now owned by just 1% of the population. So inventory management broke this further onto 10/20/70,[12] which substantially supported the production gains in U.S. manufacturing in the 1990's. For AT design, based on the gathered and imagined requirements, I have used the more approximate 80/20 split for planning implementation of complex AT. By producing robust solution to the 20% of the events and desires that end-users encounter in use of the applications and providing functions that get human help for those 80% of types of events that are only 20% of cases (see the two-basket approach in Section 3.2), it is more likely that the limited resources are utilized in the most effective manner.

The most difficult part of using this approach is to get realistic, real-world estimation of what those 20% of cases are. Domain experts are the most useful, especially for applications that are, to some degree, emulating existing work practices. Another way to approach this is to make some assumptions about the most frequent events and also the most difficult to capture/mitigate events. Then design and architecture that will accommodate both in a modular fashion categorize the estimated event and contact a human support for the more difficult one (see Section 2.2). By doing this you can test the general concept and integration of the system while making it easier for the implementation of the more difficult/rare events.

Several things can make this easier. One, error studies have shown that the absolute count of error event classes is typically a handful (see Section 4.1), and, two, the problems can be made more tractable by constraining the parameters like the COACH project did successfully.

An "inverse" example of the low-hanging fruit approach was Ritchie and Thompson's (1974) decision not to implement in kernel code provision for breaking certain kinds or quite rare deadlocks in their UNIX operating system. Their decision not to do this was the trade-off between the cost of including this (the overhead of this running would slow down the operation of all processes) and the frequency of the event occurring (I can't find a reference to this but in my undergraduate OS class it was estimated to be something like every 100,000 hours).

4.2.1   CANONICAL PAPER

Interestingly enough, J.M Juran the evangelist for quality control and management consultant, later recanted his widespread promulgation of Pareto's assumption for industrial management (Juran, 1975). However, before he did so, use of it influenced the practices of many practitioners of materials and production management in the U.S. and elsewhere.[13]

Juran, J.M. (1950). "Pareto. Lorenz, Cournot Bernoulli, Juran and others" (Juran, 1950).

---

[12]   In materials management, the ABC analysis (or Selective Inventory Control) is an inventory categorization technique. ABC analysis divides an inventory into three categories- "A items" with very tight control and accurate records, "B items" with less tightly controlled and good records, and "C items" with the simplest controls possible and minimal records. The selecting function is typically total yearly cost and loss cost due to missing inventory.

[13]   http://www.apics.org/

4.2.2   AT EXAMPLE

The principle that kept MAPS from becoming an intractably difficult problem involved reconsideration of the original goal: to support independence for a realistically large population, a subset of all of the people with cognitive disabilities. Domain experts considered "low-hanging fruit" of this subset of the population to be large enough (Braddock 2006). This subset population needed to be bound further by the limits of "could use the system" and "needed to use the system" (see Section 2.7). Given that, an existence proof could be made to show that it was possible to create scripts that would guide the user to do real (not trivial) tasks for ADL support. Assumptions had to be made about partial order plans and the resilience of this band of the population of people with cognitive disabilities as a basis for the design of such a system.

ASSISTANT Pedestrian Guidance

In the ASSISTANT project, preliminary trials of using smartphone pedestrian GPS were a failure (see Figure 4.3). There are many personal and research papers on the urban canyon effect (Martinek and emlicka, 2010), and antenna size that affirm this. The reason I think that so many projects and products rely on this technology is due to the success of the automobile systems on which so many drivers rely. One of the reasons that the automobile applications work is that they can make assumptions that if you are anywhere on or sometimes near the road it snaps the display to the road. However, pedestrians are not so predictable and due to their size need tighter granularity of location. GPS pedestrian navigation guidance has been quite successful in non-urban environments, due to the lack of the aforementioned urban canyon effect.

Red, Reittiopas
Blue, Google

Figure 4.3: Actual urban route taken vs. smartphone GPS. From VTT (2013).

In an earlier project, HAPTIMAP (2012a), that Tecnalia worked on, there were similar problems. The toolkit that HAPTIMAP project (2012b) provided for smartphones supported traveling a route also. With the toolkit HAPTIMAP consortium members made applications that successfully guided typically able users to locate each other at huge open air music festivals and hike on trails well outside of the city. However, testing the very well done interface for blind and blind-deaf users in Madrid's city streets required using a partner with a smartphone "spoofing" the GPS coordinates. It was a little better in the city park but still smartphone GPS was unusable.

Given that ASSISTANT's specification for the system, when it was funded (ASSISTANT, 2012a), had the requirement of providing guidance for the "last kilometre" from the terminal vehicle stop to the door of the route goal this could not be just abandoned without further exploration. At a technical meeting with the consortium members we explored the boundaries of what was reliable about both the system state and the GPS signal. We knew when they get off the bus, where they were, and where the goal was, so an appropriately scaled and easy-to-use level of detail map of the walking route was provided. Also, averaging the GPS signal over time and using the built-in compass in the smartphone, the absolute direction of the goal was known and the system provided a compass arrow for the direction they should be going. This took the users a bit of getting used to because the arrow pointed directly at the goal (i.e., through buildings, and across the middle of streets); but doing this was modeled on the answer to the common question asked of others at the bus stop—"Where is the Brill Building?"—and the success that resulted with a finger point of just the "as the crow flies" direction. Finally, we provided a warning and compass if the user had walked in the wrong direction more than could be accounted for by zig-zagging through waypoints. This final support was a bit tricky because of the end-user tendency to abandon applications that provided too many false positives (type 1 errors).

The resultant tests on the application (13 users in 3 cities for 1 week with no support other than an emergency phone) were successful (ASSISTANT Consortium, 2015).

4.2.3 CONCLUSION

The concept of first doing what can be done can be a very usefull strategy for finding out really what the answer to the problem will resemble. However it's very important to make sure that in doing this (which may be the only way to make a wicked problem more tractable) that you are not painting yourself into a corner, and also making maintenance a nightmare. Draw the big picture and if low-hanging fruit is applicable, make sure that your solution is malleable to suite the eventual larger system. Sometimes the low-hanging fruit is all you need; sometimes it's impossible to solve just the easy bits without a bigger covering solution. Like programming problems, design work, at its most effective, is 70% thinking and planning then the 30% of actual work.

4.3   METADESIGN

Short Definition: MetaDesign is a theoretical framework for supporting open systems to allow end-users (and other stakeholders) to become designers in dynamic use contexts.

Longer Description: Of all the concepts discussed here, MetaDesign is perhaps the most difficult to pin down. This is for several reasons. First, it is used in pretty specific ways in disparate domains that do not completely overlap. So in the original context of Industrial Design MetaDesign is quite different in description and action than in Maturana and Varela's biological approach (Maturana and Varela, 1987; Maturana, 1997), and Giaccardi's and Fischer's socio-technical approach (Fischer and Giaccardi, 2006). Additionally, there are a variety of ways to even spell MetaDesign: MetaDesign, Meta-design, Meta Design. In this book, I will use the spelling MetaDesign.

In task support AT tools and more generalized AT/IA applications MetaDesign is a way to talk about how to design tools that allow dyads (see Section 3.1) to together create and use supports for doing that which the end-user cannot do by themselves. This is a version of what Fischer and Giaccardi discuss as design for designers. This design approach refers to making artifacts that can be used to design further artifacts. To make sense of this, one needs to break the end-users into two groups, the (secondary) tool maker, in most cases a caregiver or more technically savvy member of the dyad, and the end-user herself, the person who used the created tool. There is a third person in the MetaDesign constellation, the MetaDesign tool designer/creator. Table 4.2 illustrates this for the example of the MAPS system.

Table 4.2: Overview of MetaDesign attributes for MAPS

Stage	Initiator	Recipient	Artifact
Tool-design time	Tool designer	Caregiver	The script editor
Artifact–design time	Caregiver	Person with cognitive disabilities	Scripts and re-designed scripts
Artifact-use time	Person with cognitive disabilities	Caregiver	Script use logs and observed behavior

Here is the same table for the ASSISTANT project.

Table 4.3: Overview of MetaDesign attributes for MAPS

Stage	Initiator	Recipient	Artifact
Tool-design time	Route creator designer	End-user, could be caregiver	The web-based route editor
Artifact–design time	Elder/caregiver	Elder/caregiver	Route for PND (can be re-used)
Artifact-use time	Elder	ASSISTANT server monitoring	Output new route or SMS to help contact if appropriate

Among the attributes of this kind of MetaDesign for AT that are relevant to AT designers are the different artifacts to be generated, the target user population, and the need for and danger of underdesign. Typically a MetaDesign project has at least two interfaces and one shared database (or persistent data of some sort). The first interface is for the person who will configure the final tool for task support; in this is a response to the universe of one problem and, in some ways, an ultimate personalization tool. The second interface is for the end user that will use it as support for task completion. The first tool will often not just create content to be played on the end-users interface but also may produce configuration commands for the second tool's user interface. In this way the first tool becomes an example of end user programming (Nardi, 1993; Lieberman et al., 2006), and as such the designer may well profit from studying work on this topic.

Underdesign of the first tool needs to be guided by the position on the axis of, at one end, only supporting a very limited set of results to be used in the second tool, and on the other end of the continuum, producing a very general first tool but on that takes deep programming skills to do anything. A MetaDesign system at the first end may be very easy to use for the initial intended audience, but does not generalize well; at the other end it may be completely unworkable due to complexity. This is another example of the Turing tarpit (Perlis, 1982) where the tool may be able to do anything but anything of interest is impossible to do.

One solution to the underdesign problem is to clearly understand the internal scripts and skills of the intended end users, and to constrain the solution domain tightly enough so that appropriate support elements for that domain are built into the first tool.

4.3.1 CANONICAL PAPER

There are many relevant publications on MetaDesign (which can be seen in Part 3), however the paper below represents the seminal use of MetaDesign in this domain. While not specifically pointed at AT it does capture all the elements that became the attributes in MetaDesign for AT.

Fischer, G. and E. Scharff (2000). "Meta-design—Design for designers." (Fischer and Scharff, 2000).

4.3.2 AT EXAMPLE

The section above on describing and designing MetaDesign in the MAPS system has said much about an AT example, there are some other dimensions worthy of discussion. First is the issue of MetaDesign over time, second is the idea of reusable and personalized components.

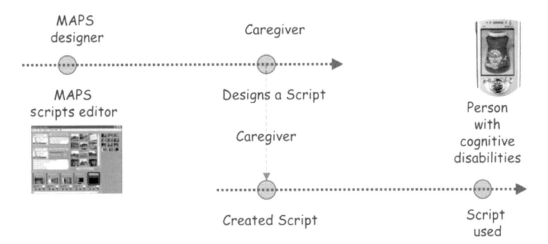

Figure 4.4: **MAPS over time.** From Carmien (2006b).

There are two timelines in MAPS MetaDesign, the design process, and the use practice (Figure 4.4). The MAPS system designer actually designed prototypes (corresponding to wicked problem iteration) of the MAPS script editor, hand-held prompter, and the database schema and functionality to bridge data between them. Once several iterations were tested and refined the realistic testing were done in a time span related to system use. One of the most difficult and important parts of the first tool, the scripts editor, was making it easy for the caregiver to use digital photographs that they had taken for the script; although the system could accommodate icons, that were provided, almost all the images were personal images collected by the caregiver. Similar care was taken with the recorded verbal prompts. Both of these were stored in a database rather than taking the easier route of using the Microsoft file systems and something like XML (as AbleLink's commercial offering (AbleLink, 2002, 2007) and proposed standard (Technologies does) to ensure atomicity and robust reliability in an already fragile environment (i.e., IPAQ running winCE and windows in 2004).

4.3.3 CONCLUSION

Ignoring the need for a MetaDesign approach to system design can lead to a lack of ecological validity of the solution. I have seen quite a few projects and proposals that thought a lot about the end user's artifacts, and did not pay enough attention to the creation of the content in the artifact. Examples include work on learning material accessibility systems (see the discussion od EU4ALL) and scripting systems for the IoT, how are the scripts created and edited?

4.4 PERSONALIZATION

Short Definition: Personalization is the process of structuring a system, both the user interface and the systems functionalities, to more precisely fit the abilities, needs, and desires of the end-user.

Longer Description: Practically every user interface since the Teletype has had some facilities for customization, from terminals where you could invert the screen background and foreground, to changing the voice accent in text to speech interfaces. The axis of customization runs from convenience to necessity. I can change the font type in my browser, which is nice. But my almost blind son-in-law needs a magnified screen to read his assigned text in engineering school. Here I am discussing the need end of the continuum.

The literature on personalization typically focuses on automatic interface personalization (Mourlas and Germanakos, 2009), and the first example below shows how, laboriously, such automatic service works, as well as the hidden drawbacks that are not often discussed. I have gone through the personalization workflow in (excruciating) detail not so much to provide a model to follow but as an example of how complex and interconnected doing adaptation can be in the practice. Personalization of the interface by providing the tools for a caregiver or end-user to configure their user environment has been the bulk of my work.

To simplify the issues involved in automatic and self controlled personalization I introduce the notion of the two kinds of personalization, Ser and Estar personalization Models. Whether explicit or tacit, behind all personalization systems lies a model of the user. This abstraction allows computational personalization to be implemented.

One common design problem encountered by user model systems is not distinguishing between user model attributes that represent the users "being" and the users state. This leads to unnecessary data duplication and can migrate up to the user interface, causing confusion in the system's end-user and unnecessary duplication in the system's architecture and persistent storage. Looking at the whole of user modeling data, it basically falls into two kinds of models, which I have called the ser and estar types of user models.

I have chosen to use two Spanish verbs that express essential and acquired qualities because their use roughly corresponds to the two kinds of data: one uses *Ser* to refer to the essential characteristics of things that you are, your name, your gender; *Estar* is used to refer to descriptive, potentially temporary attributes, the condition, like *I am walking*. The Ser user model contains the information about the users—these essential characteristics form the static user model. In Ser models changes occur very infrequently over time and they are typically independent of context. The Estar user models, in contrast, contain changing values that model the current condition of user; this dynamic user model may change over time, or there may be multiple copies that are linked one at a time to the ser model of the user. Estar models are often context dependent or reflect the different devices that are used at different times. Estar data type is used to keep histories of user actions.

This novel user model framework is driven by a relational database approach to users and contexts (i.e., normalization (Codd, 1990)), and inspired by several experiences with projects and models where the two were conflated, with problematic consequences. By keeping data that never changed in the same "table" as attribute values that change frequently, the system is forced to either keep multiple copies of a user model for a single person, or to constantly change the model as the context of use changes. Also, conflation of the two types makes keeping histories complicated and causes redundant data to be stored. This section will discuss the ramifications of these approaches. This is not a "new" approach per se, but by explicitly creating a framework that holds these two models design and performance advantages can be accrued.

Another advantage of discriminating between ser and estar models is that this provides a nice way to separate the actual data, so that applications with large amounts of sensor data can store it locally, making a mobile system less dependent on reliable high bandwidth connectivity.

Looking at these two model types, most of the effort to do automatic personalizing should be applied to estar data. And, in most cases, the Ser model data can be quite simply setup once and not need further attention. It is these Ser preferences that are on the preferences page of ASSISTANT (see Figure 4.5).

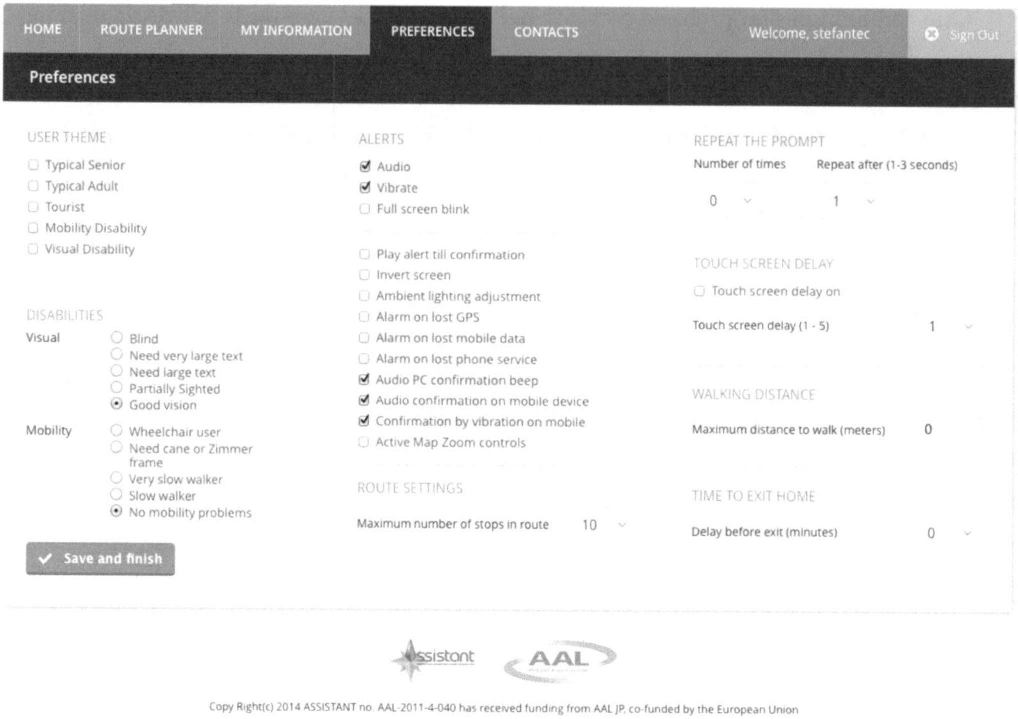

Figure 4.5: Preferences choices in ASSISTANT. From Assistant Project (2015).

4.4.1 CANONICAL PAPERS

There really are not any "canonical" publications about personalization, especially not for static personalization. Below is a good start at dynamic personalization overview, and the human interface apple guidelines do a good job of taking about the interface for the end-user.

Mourlas, C. and P. Germanakos (2009). "Intelligent user interfaces: Adaptation and personalization systems and technologies." (Mourlas and Germanakos, 2009).

Apple, Inc. (2009). *Apple Human Interface Guidelines* (Apple, 2009).

4.4.2 AT DESIGN EXAMPLES

A current topic in European research is systems supporting universal access to distance learning,[14], [15,16] the specifics described here came out of initial design work in the EU4ALL project (EU4ALL, 2007). These projects emphasize the use of content fitted to stereotypical sensory, and to a lesser

---

[14] http://www.aegis-project.eu.
[15] http://adenu.ia.uned.es/alpe/index.tcl.
[16] http://www.eu4all-project.eu/.

degree, cognitive disabilities. Adding to the complexity of the problem is the rising popularity of mobile platforms; where previously a reasonable assumption was that the material would be accessed through "typical" PCs, now material may be presented on screens as small as a smartphone and in the pad format. The rise of ubiquitous and embodied computational services adds to the complexity of delivery of the right content, in the right way, at the right time, and via the right medium (Fischer, 2001b).

This kind of personalization is a special case of a broad range of applications and systems that will be tailored to a user's needs and capabilities and with respect to the context at the time and place of use. Scenarios of use may include searching and accessing schedules and real-time location of buses in transit services non-visually on a smartphone (Sullivan, 2004); reading medical records on a wall display closest to the users current location and printing them on the nearby printer; locating a friend in a crowded shopping district, locating a film center nearby and purchasing tickets, and using a mobile pad to present a text version of a lecture given as part of a distance learning university course.

This kind of deep personalization is dependent on three things: the end user, the context (which includes the device used for input and output), and the content. Each end-user presents a unique set of abilities and needs; sometimes this may be as broad as a preferred language or as deep as sensory and motoric disabilities with preferred alterative presentation mode (e.g., low visual acuity and synthesized text) and preferred input mode (i.e., voice commands, scan and select input). The context of the application has two dimensions: the actual environment, including the history of the user's interactions with the computer (in general) and earlier sessions with this application; and the device with which the user interacts with the system, including the device constraints (i.e., small screened smartphones) and the device capabilities (i.e., speakers, text synthesizer). The third part of the personalization is the content that is presented to the user; this may be one of a set of identical "content" expressed in different modalities, or a server that adapts the content to the needs and capabilities of the user/device.

It is difficult to discuss a system that uses user models without considering the context in which the user models provide leverage. Start by considering the user herself, the User Model (UM). While the specifics discussed here are concerned with the needs and abilities that differ from "typically" abled users, looking at the problem with this lens makes it easier to highlight the salient issues for all users. Each person brings to the problem space a unique set of abilities and needs, and it is important to consider both the disability *and* the preferred mode of adaptation that this user has. Because of the unique set of attributes that the user brings to the application the use of simple stereotypes, while initially appealing, may cause a poor fit (due to the too large granularity of the stereotype) between document, applications and user. For these reasons these user models have attributes that discretely describe individuals to the adaptation system. Examples of these discrete attributes may include various levels of visual disability, where describing their visual sense with a binary blind/not blind may not capture the large percentage of partially sighted people who could gain some advantage from a custom form of visual interface. Similarly, motoric and cognitive disabilities both require a wide range of attribute-value pairs to describe them. Further, a given person may have a *combination* of sensory/motoric/cognitive abilities that makes them unique. Ontologies and hierarchical description schemas may be useful, both as a support for attribute names but also in that nodes that are not leafs may also provide goals for places to base accessibility support. Finally, this part of the user modeling system should only concern itself with those attributed to the user, which never change, or change very slowly monotonically down an axis of ability. This means that if a user condition pertinent to the application waxes and wanes, in contrast to only decreasing over time, this should be captured as a part of the context /device model as described below.

The next element in this user modeling system is the device model (DM), which in a larger sense is the context. This means several things: (1) the actual context which includes the current time and place and various other details that constitute the environment; (2) the device used to access the material; and (3) the changing attributes of the user. To describe the actual physical context existing ontologies and frameworks may prove both useful and a way to access existing data sources. Part of the captured context includes resources as well as static descriptions, i.e., network accessibility, temperature, light.

Device as context makes sense if you talk about context as everything on the outside of the user that affects, or is pertinent to, the user. In the case of the user's device this provides an inventory of capabilities to the system. The inventory of device abilities can include local availability of resources (i.e., printers, Java, browser) I/O capabilities such as a list of compatible mime types (IANA (Internet Assigned Numbers Authority), 1996) and input affordances (touch screen, speech synthesizer, and verbal recognition). These qualities can be represented in existing schemas (frameworks or ontologies) such as UAProf and CC/PP (Open Mobile Alliance, 2007; W3C, 2007). A more dynamic approach for device description can be taken with systems like the examples of V2 (trace.wisc.edu, 2005) and URC (URC Consortium, 2007), or the Aria initiative (W3C, 2009a) where the device can self configure to match the needs of the applications user controls.

Finally, context can be considered as the history of the user and application. History as context can be built from system history, application history, and user's history (as captured for example, in bookmarks and previous preferred configurations). The user modeling system can use these as both local support for making inferences, i.e., he is doing this task and has just finished this subtask, and as a basis for inferring conditions of importance to the successful completion of the goal-at-hand. An example of this could come from a task support application (Carmien, 2007) where the user may have paused for a long time at this point in previous trials (and might trigger additional help from the system) or where the user is going over the sub task prompts very rapidly, from which the application may infer that the user has these sub-tasks Memorized in a chunk and therefore might flag compression of many prompts for a set of sub-tasks into one single prompt for a larger sub-task.

The last part of this user modeling system is the content itself. It may be easier to think of the content as a participant in the accessibility process to talk about it in process terms. This example uses the term Digital Resource Description (DRD), a set of metadata associated with documents. In the educational example being discussed every document or "chunk" of digital material in an educational process has a DRD record. This approach comes out of the work that has been done in learning object metadata (LOM—IEEE 1484.12.1-2002 Standard for Learning Object Metadata) (IMS Global Learning Consortium, 2007). A LOM is a data model, usually encoded in XML, used to describe a learning object (a chunk of content). In this case the specifics were drawn from a pre-release copy of the ISO 24751 standard (ISO/IEC 2007a, ISO/IEC 2007b, ISO/IEC, 2007c).

The provision of appropriate material results from this set of attributes and values fitting together with the needs of the user as expressed in her user model and the capabilities of the user current device as express in the DM. This approach can take two forms: (1) selection and delivery, and (2) adaption. Both of these have network bandwidth requirements for successful use, which are part of the information on the DRD record. The selection and delivery approach finds the existing right content based on the user and device/context and presents this to the user. The adaption approach takes a Meta document representing the content and creates the accessible content on the fly.

Select and deliver requires the system to create and store different versions of the same material, and this leads to authoring and on-going revision problems. The advantage of replacement it that it is relatively easy to implement, after the matching infrastructure in the UM and DM are in place. The problem with this approach is that content authors have the burden of many adaptations and later modifications of material will make sets of the same content out of synch. Adaptation forces the author to write in a kind of markup language, which may be mitigated by authoring support tools. Adaptation also requires another layer of software to take the marked up content as input and produce the appropriate content; this may either be done on the server or at the client. Also adaptation may support interface adaptation like ARIA (W3C, 2009a) and URC/V2 (INCTS, 2007), taking motoric issues in accessibility into account.

Adaption, while a complete solution, requires standardizing and built-in functionality like cascading style sheets (CSS), which have been used in this fashion with some success. Select and deliver can use existing systems but adaptation requires a whole new infrastructure, and more relevantly, the adoption of this infrastructure by all content developers.

Having described all the parts of the user modeling for accessibility system, it can be now described it in shorthand as:

UM + DM + DRD = CP, which can be called the *Content Personalization* formula.

Where UM is the user model. DM refers here to the Device Model, remembering that it includes context and history in the DM. The DRD refers to the metadata that describes the content at hand, which may take many different forms while pointing to the same concepts. Finally the CP stands for content personalized for the user and context. In order for this to work there must exist a tightly controlled vocabulary, which is often domain specific, so that UM, DM, and DRD must have matching attributes and ranges of values. It is important to ensure adoption of such systems by building on existing standards, such as shown in Table 4.4.

Table 4.4: Existing personalized distance learning standards
LOM (IEEE)
Dublin Core (DC-education extensions)
IMS
ACCLIP
ACCMD
ISO 24751-1,2,3
DRD, PNP

Here are existing standards used in the following example.

- **User Modeling (UM):** PNP (personal needs and preferences from part 2 ISO 24751 draft 2007).

- **Device modeling (DM):** CC/PP (from W3C Composite Capabilities/Preference Profiles: Structure and Vocabularies 2.0 (CC/PP 2.0).

- **Content Metadata:** DRD (from Part 3: Access for All ISO 24751 draft 2007).

So the content personalization formula then becomes:

PNP + CC/PP + DRD = CP

Here is an example of the process, in our prototypical remote learning accessibility system (Figure 4.6). First, there is a request for content object (CO) in learning process. To do this, the user agent (browser) uses a proxy that inserts device model ID into the header of http request containing the request for a CO in the form of the device identifier. Then the virtual learning environment (VLE) passes these (the content ID, the UM-ID, and the DM-ID) to a content personalization module (using web service calls from here to the return to the VLE). Content personalization module then gets:

1. user profile from user modeling subsystem;

2. device profile from device modeling system;

3. CO accessibility metadata-digital resource description (DRD) from Learning Object Metadata Repository; and

4. it matches them and returns the right one (if there is one).

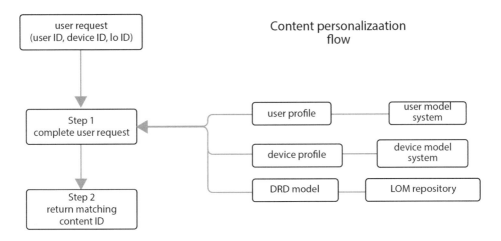

Figure 4.6: Content personalization flow in select and deliver model. From Carmien and Cantara (2012).

An alternative process flow, using the adaptation model is seen in Figure 4.7.

Following that are examples of both the select and deliver approach and the adapt approach; to deliver accessible educational materials to students, matched to their needs and the capabilities of the device that they are using. To effect this the three parts of the CP equation need to be considered. This can be expressed as shown in Table 4.5.

This process is further broken down in Tables 4.5–4.13. Table 4.6 showing the Adaption Preference from PNP; Table 4.7 containing Preferences from PNP; Table 4.8 giving an example of DM Attributes from CC/PP; Table 4.9 laying out the parts of a Media Meta Record from DRM; Table 4.10 explicating how Table 4.6 is comprised of the Access Mode Record Template; Table 4.11 expands on Table 4.10 into an Adaptation Template; Table 4.12 illustrates the use of the Adaptation Template; and finally Table 4.13 shows the whole process in the Content Selection Matrix.

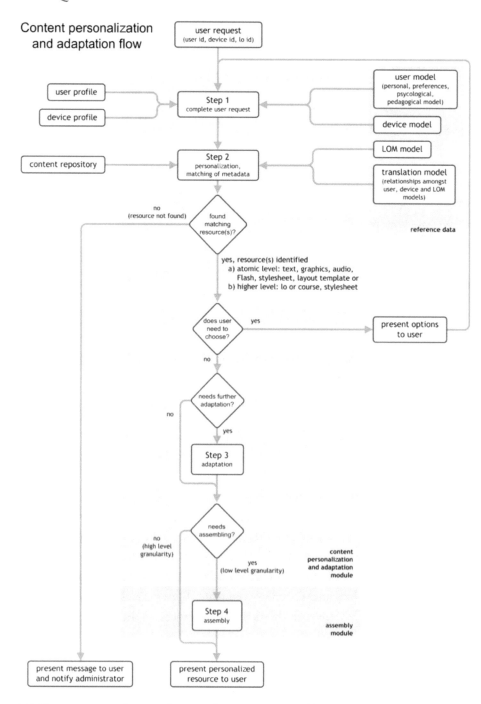

Figure 4.7: Content personalization flow in adaptation process. From Carmien and Cantara (2012).

Table 4.5: Three parts of the CP equation

No.	Table	Example	Orig. access mode (s)	Adaptation type
1a	Auditory	Tape of talk	Audio	Transcription
1b	Visual	Text of lesson	Text	Audio tape
2	Visual	Physics lecture on video tape	Visual and Auditory—Two entries in metadata pointing at the same original content object: visual and auditory	
2_{visual}	Visual	Demonstration part of above	Visual part	Audio description
$2_{auditory}$	Visual	Lecture part of video	Auditory part	<none> NOTE: this is an *.avi
3	Visual	Photo caption	Visual	OCR
4	Visual	Text (a book)	Visual	Text to audio DM transformation

The PNP record in the user model stores the user's preferences of content presentation. The index into the preferences table:

Table 4.6: Adaption preference from PNP

Attribute	Allowed Occurrences	Datatype
Adaptation preference	Zero or more per Content	Adaptation_Preference

The preferences table itself.

Table 4.7: Preferences from PNP

Attribute	Allowed Occurrences	Datatype
Usage	Zero or one per adaptation preference	Usage_vocabulary
Adaptation type	Zero or one per adaptation preference	Adaptation_type_vocabulary
Original access mode	One per adaptation preference	access_mode_vocabulary
Adaptation_preference _ranking = not DM or DRD	Zero or more per adaptation preference (i.e., multiple adaptation types for the same original access mode could exist)	Integer—the preference rank of the possible adaptation

Having gotten the user needs from the above, the personalizing process continues with the Device Model (in this simple case there are no context or historical data, but this is where they would go). Here the DM is based on the CC/PP record from UAProf (Wireless Application Protocol Forum, 2001):

Table 4.8: DM attributes from CC/PP		
Attribute	**Description**	**Sample Values**
Mime_Type	List of the IANA mime type(s) that can be "played" on this device	"Audio.MP3" See IANA mime type listings for vocabulary
AT-Transformation type	A bag of literal strings, each literal represents a given transformation (scenarios)	Could be integers could be audio-to-text

Where the mime types are:

- .aif audio/aiff;

- .aos application/x-nokia-9000;

- .asm text/x-asm;

- .eps application/postscript;

- .gif image/gif;

- .html text/html;

- .java text/plain;

- .jpg image/jpeg;

and the transformation type could be:

- Text -> audio;

- Audio-> text; and

- HTML-> correct color contrast.

Finally, there is the metadata attached to each chunk of content. This is taken from the ISO 24751 Part 3, the DRD record template. The media META record:

Table 4.9: Media meta record from DRM

Attribute	Allowed Occurrences	Datatype
Media_object_id	One time per Access For All Resource	EU4ALL Identifier
Access mode statement	Zero or more per Access For All Resource	Access_Mode_Statement
Has adaptation	Zero or more per Access For All Resource	EU4ALL Identifier
Is adaptation	Zero or one per Access For All Resource	Is_Adaptation
Adaptation statement	Zero or more per Access For All Resource	Adaptation_Statement
Mime_type—DM	Zero or more per Access Mode Statement	IANA Mime type

An access mode record:

Table 4.10: Access mode record template

Attribute	Allowed Occurrences	Datatype
Original access	One per Access Mode Statement	access_mode_vocabulary
Access mode usage	Zero or one per Access Mode Statement	access_mode_usage_vocabulary

An "is adaptation" record:

Table 4.11: Is adaptation template

Attribute	Allowed Occurrences	Datatype
Is adaptation of	One per Is Adaptation	EU4ALL Identifier
Extent	One per Is Adaptation –	extent_vocabulary

And finally, the adaptation statement (AS):

Table 4.12: Adaptation		
Attribute	**Allowed Occurrences**	**Datatype**
Adaptation type	Zero or one per AS	adaptation_type_vocabulary
Original access mode	One per AS	access_mode_vocabulary
Extent	Zero or one per AS	extent_vocabulary
Mime_type	Zero or more per AS	Mime type vocabulary from IANA site

Putting them altogether:

Table 4.13: Content selection matrix					
Disability	**Original access mode**	**Adaptation type**	**PNP attribute / values**	**DRD Attribute / values**	**DM Attribute / values**
Needs replacement for auditory	Audio	Transcription	Orig. access. Mode = Auditory Adaptation type = Text represen-tation // Adapta-tion_ preference _ranking = 1	Adaptation stanza: Orig. access. Mode = Auditory // Adaptation type = Text represen-tation // Mime_ type = Text.plain	Mime_type = text.plain
Needs replacement for visual	Text	Audio tape	Orig. access. Mode = Text Adaptation type = Audio repre-sentation //Ad-aptation_ prefer-ence _ranking = 1	Orig. access. Mode = Text Adaptation type = Audio repre-sentation Mime_type = Audio.MP3	Mime_type= Audio.MP3

Here is a more complex example requiring multiple adaptations (Figure 4.8). In this example there is a video of a physics lecture and a demonstration. The video has two parts: audio and visual.

Disability	Example	Orig. access mode (s)	Adaptation type	PNP attribute / values	DRD Attribute / values	DM Attribute / values
Visual impairment	Physics lecture on video	Visual and Auditory Two entries in metadata pointing at the same original content object: Visual and Auditory				
Demonstration part of above		Visual part	Audio Description	Orig. access. Mode = Visual Adaptation type = Audio_Description Adaptation_ preference _ranking = 1	In the adaptation stanza Orig. access. Mode = Visual Adaptation type = Audio description Mime_type = audio.mp3	Mime_type = audio.mp3
Lecture part of video		Auditory part	<none> NOTE: this is an *.avi	Orig. access. Mode = Auditory Adaptation type = <no entry>	Mime_type = video.avi (we are only dealing with the audio side of the avi)	Mime_type = video.avi

Figure 4.8: A more complex example of selecting content. From Carmien and Cantara (2012).

Finally, an example (Table 4.14) using device-based transformation.

Table 4.14: Content transformation example	
Disability	Visually impaired
Example	Text (a book, etc.)
Orig. access mode	Screen reader
Adaptation type	Orig. access. Mode = Text Adaptation type = Audio representation Adaptation_ preference _ranking = 1
PNP attribute/values	In the adaptation stanza Orig. access. Mode = Text No matching adaptation type
DRD Attribute/values	In the adaptation stanza Orig. access. Mode = Text No matching adaptation type
DM Attribute/values	DM has transformation attribute that maps from text to audio (i.e., has jaws or the like)

The above brief introduction to the complex domain of personalized content delivery is but one of many possible implementations of the UM + DM + DRD = CP formula. In the author's opinion, the content selection and delivery approach is seriously flawed due to the authoring and updating requirement. By this I mean that although a given "chunk" of content may exist in many different media, covering most possible needs and disabilities, any updating of the content requires simultaneous updating of all variations, a condition that has a very low probability of happening. On the other hand, the adaptation approach, while complete, requires the adoption of a meta-language and adaption engines across the domain, which requires an integrated system to be designed, implemented, and widely adopted. Because of the difficulty of doing this current efforts are typically focused on the select and deliver approach.

The other approach widely implemented is writing educational content in a markup language. The wide implementation is, of course, the World Wide Web. A browser can take as in input an html/css file set and produce very different output depending on the settings and add-ons in the browser. The drawbacks to this approach are two-fold:

1. All education content authors and consumers must agree on the markup language and the produced file interpreter. This worked well for the WWW and there is work being done in the W3 consortium (W3C, 2009b), but the LOM community has demonstrated needs that HTML/CSS may not apply to.

2. Education content authors now must either master the markup language or authoring tools need to be implemented and widely accepted.

So this automatic problem is very real and not solved in a fashion amenable to wide adoption. Either endless updates of many, many different expressions of the same content, including the problems with orphan files and enslaved authors; or making all content to be written in a markup language with near universal adoption of a markup language reader.

4.4.3 CONCLUSION

So, there are two types of personalization: (1) static that can be fairly easily implemented and is mostly dependent on the perception of the needs of the various end-users and proposed functionality of the application, and (2) dynamic that reflects the changing context of the user in action. It is important not to confuse static and dynamic; the conflation can cause both problems with data and determining which user is the user. It is important that, as much as possible, these static preferences are scrutible, that the end-user or caregiver can see and change them as needed. The second type, personalized dynamic content delivery, is much more complex (see the length of the illustration above) and has non-trivial problems with authoring tools and updating.

Some very promising work is being done in dynamic personalization and content representation in the Global Public Inclusive Infrastructure (GPII)[17] consortium and the work of Gregg

---

[17] http://gpii.net/About.html.

Vanderheiden and Jutta Treviranus. This is the current result of a long process stating in the URC (trace.wisc.edu, 2005; Zimmermann et al., 2006) work in the 1990s leading to the Raising the Floor initiative.[18] This group has done some very pertinent work ion addressing the "this will solve the problem perfectly, BUT everyone must adopt XXX framework/language/notation" dilemma. As of 2016 this is a project to keep an eye on, and perhaps participate in.

4.5   SYMMETRY OF IGNORANCE

Short Definition: Since the Renaissance it has not been possible for any one person, no matter how talented or educated, to have all the knowledge required for design or creating a non-trivial system.

Longer Description: The symmetry of ignorance (or asymmetry of knowledge) is a way of describing situations in which several participants or roles in an endeavour each individually have parts of the knowledge needed to accomplish the task, but none has enough to accomplish the task independently. Symmetry of ignorance or asymmetry of knowledge (Ostwald, 2003) refers to the fact that for many "real-world" problems no one person or role has all the knowledge to design a satisficing solution. A satisficing solution (Simon, 1996) refers to a solution that is "good enough" for all the stakeholders to achieve some minimum of desired functionality. This is a useful notion, because any attempt to reach an optimal solution is very difficult due to limits on resources and time.

As Gerhard Fischer pointed out (Fischer, 1999):

> *The Renaissance scholar does not exist anymore. Human beings have a bounded ratio-nality—making satisfying instead of optimizing a necessity. There is only so much we can remember and there is only so much we can learn. Talented people require approx-imately a decade to reach top professional proficiency. When a domain reaches a point where the knowledge for skillful professional practice cannot be acquired in a decade, specialization will increase, collaboration will become a necessity, and practitioners will make increasing use of reference aids, such as printed and computational media support-ing external cognition.*

> *Much of our intelligence and creativity results from the collective memory of commu-nities of practice and of the artifacts and technology surrounding them. Though creative individuals are often thought of working in isolation, the role of interaction and collab-oration with other individuals is critical. Creative activity grows out of the relationship between an individual and the world of his or her work, and out of the ties between an individual and other human beings. The basic capacities … are then differentially orga-nized and elaborated into complex systems of higher psychological functions, depending*

---

[18]   http://raisingthefloor.org/who-we-are/our-approach/.

on the actual activities in which people engage. These activities depend crucially on the historical and cultural circumstances in which people live (Resnick et al., 1991).

There are two primary components to this need for a community of interest (Wenger, 1998) approach to larger design and implementation projects. First, is the mastery of skills, where there is simply not enough time for any one person to obtain all the relevant skills, especially up to date at the same time. Second, is the need for many perspectives, similar to the need for stakeholder analysis (Overseas Development Administration, 1995) in the early stages of the project. Again, not enough time to gain the "observational" experience to make the design effective, and in the case of AT successfully adoptable.

How can one expect an individual to maintain the requisite specialist knowledge in their technological discipline, while at the same time have the needed competence in industrial design, sociology, anthropology, psychology, etc.? The solution to this dilemma lies in the notion of Renaissance team (Buxton, 2002): a social network of specialists from different disciplines working as a team with a common language. One way of seeing this is the exploitation of the symmetry of ignorance, by exploiting my expertise in an area you are lacking in (and vice versa, over a group). This requires support by CSCW tools and structured participation to be more than inconsequential. In a balanced team the deficiency of knowledge in one member will be balanced by the expertise being held by another member, and this symmetry is spread across the research team

An example of not attending to the need for using the perspective of symmetry of ignorance that I experienced as a MIS manager for a hotel reservation company (Express Reservations, Carmien (2002)) was shopping for a new hotel reservation system. Those we looked at that were designed by programmers alone ran and were quite robust and extendable to our needs (a phone-based reservation system based in Boulder Colorado for hotels in a 33-square block area in midtown Manhattan), but were virtually useless for making actual hotel reservations and the data needed behind the front end. The offerings designed by hotel reservationists had a totally fitting interface, supporting existing work practices (i.e., understood what a reservationist needed and how and what way to store the information about hotels, reservations and rates) that was needed and in the way that fit best—but continuously crashed

Symmetry of ignorance is part of why domain-spanning initiatives can be so successful. An acknowledgement of symmetry of ignorance is the concept of a community of interest. Communities of interest are groups of people (typically coming from different disciplines) that engage in a joint activity. Instead of being obstructed by lack of expertise one gets a synergistic coverage of the domain

4.5.1   CANONICAL PAPER

Rittel, H.W.J. (1972). "On the planning crisis: Systems analysis of the first and second generations. (Rittel, 1972)

4.5.2   AT EXAMPLES

MAPS and the role of Anja in CLever set of projects. Leveraging the Symmetry of Ignorance. There were two sets of members in the symmetry of ignorance in the MAPS system: (1) caregivers and clients; and (2) tool (AT) designers and caregivers. Here was a further iteration of the notion of symmetry of ignorance at a higher design level: the caregiver and the script editor together have the solution to an adoptable script editor, but neither has all the relevant skills and knowledge to construct an editor on their own.

Table 4.15: ASSISTANT Consortium expertise

Requirement	Partner
Knowledge of older people and technology	ESENIORS
Assistive Technology design and design for all	TECN
Knowledge of existing State of the Art	UNIVIE, TECN
System Architecture design	VTT
Route editor design	UNIVIE, FARA
Transport data analysis	VTT, FARA, UNIVIE
User modeling	TECN
Error detection	TECN
Error recovery	TECN
PND design	TECN
GIS and telematics expertise	UNIVIE
GPS	UNIVIE
System integration	VTT TECN
Field trials	ESENIORS, TECN, VTT, CIT
System validation	VTT
Commercial exploitation	TTR
Dissemination	TTR
Project management	TECN
Service Oriented Architecture	Subcontracted by TECN to IN2
Map and transportation data	Local transport authorities, mapping organizations
Develop mobile application	CITRUNA

Another example of the way that using the symmetry of ignorance can help to avoid system killing approaches. The ASSISTANT project (Table 4.15) put the way-point and map-based guidance information for elders who found exiting smartphone based transport navigation support too complex or with an interface too difficult to use. This Personal Navigation Device (PND) application was to be installed on medium sized smartphones for the targeted population. The technical partners of the ASSISTANT consortium, experts in mobile phone interface design and implementation of complex functionality at the server, phone and data supplier (i.e., schedule, routes, and real-time data from local transit companies) level, brought an enormous amount of skills and knowledge to the problem. What they did not bring to the problem was knowledge of the specific sensory and cognitive needs of the target elder population.

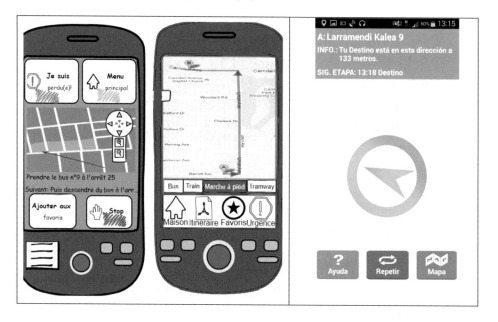

Figure 4.9: The PND GUI: left—initial designs; right—final design. From Assistant Project (2015).

Fortunately, the consortium was brought together initially for the Symmetry of Ignorance perspective, as a result other consortium members had intimate and deep knowledge of elders and elders adoption of new technology, expertise in human computer interaction with emphasis on AT design, and knowledge of research literature on navigation guidance and design of interfaces for elders. So, in the end the interface for the phone was well accepted and there were no requests for UI change by the elders after a week long trial in "the wild." However, there were many detailed and almost acrimonious discussion about the elements to include or put focus on (Figure 4.9). The positive result, besides the PND itself, was a meta study (Carmien and Garzo, 2014) of appropriate design guidelines for these types of systems, motivated by a need to demonstrate that the GUI design for elders is quite different from GUI design for 20-35 year-old programmers.

4.5.3   CONCLUSION

Taking the Symmetry of Ignorance approach may be somewhat tricky due to power issues: who has the vision and who decides the relative priority of different requirements that need to be coordinated in a satisficing solution. In ASSISTANT the solution was a well-defined set of UI specifications, general enough so that the implementation was not forced to use specific elements and controls. Additionally face-to-face meetings allow everyone's ideas to be expressed and rationale for each part to be agreed on, something that just exchanging images of deign alone does not support well.

CHAPTER 5

Things to Avoid

This brief chapter contains some dead-end warnings that novice AT developers may face. This is very much a personal list, but others may be found in Kings excellent book on AT design (King, 1999):

- Diagnosis and Functionality

- I Have a Theory; I Have a Cousin

- Islands of Ability

5.1 DIAGNOSIS AND FUNCTIONALITY

Short Definition: Many technologists, in initial forays into the domain of AT design, focus on the diagnosis resulting in the disability rather than the actual missing/low functioning abilities. This leads to many blind alleys, and missing many design opportunities.

Longer Description: As designers, we are always looking for more information about the problem at hand, and the first offered description of a potential user is typically the diagnosis. For instance, a colleague's father-in-law, a former tennis instructor, uses a wheelchair daily and when queried about his disability, the answer is that he suffers form Parkinson's. Similarly, people with cognitive disabilities are often described as having Down's syndrome, Fragile X syndrome, fetal alcohol spectrum disorder, etc., based on the etiology of the intellectual disability that they are suffering.

Ironically, terms used to describe cognitive or intellectual disabilities are subject to the "euphemism treadmill," where whatever term is chosen for a given condition eventually becomes perceived as an insult. The terms "mental retardation" and "mentally retarded" were invented in the middle of the 20th century to replace the previous set of terms, which were deemed to have become offensive. However, these newer terms have also become seen as offensive (e.g., "You retard!"). Interestingly, these terms describe the effects of the diagnosis and, as a result, systems focus on the cause of the functional disability, rather than (a very broad) description of the problem-at-hand.

5.1.1 DESIGN WITH DIAGNOSIS

Choosing to base the design requirements on a target population defined by diagnosis is, with a few exceptions, often as a result of selecting a range of functional ability that is too broad. At the same time, focusing on the diagnosis may exclude others with implicit functional needs that may prove very useful. One of the few examples of designing AT for people with cognitive disabilities using

diagnosis is designing for those suffering from Autism spectrum symptoms such that this population is often fascinated with issues of control and patterning, an issue not as much at the forefront
of requirements for those with intellectual disabilities that do not have this particular emotional
and motivational component. In the end, using the diagnosis as a basis for (most) design of AT for
IA really does not provide much leverage in the design process beside selecting a set of end-users.

Design by Functionality

By looking at functional needs the target group expands and thus, ironically, focuses the design
on ameliorating the missing ability rather than on the cause of the disability. Imagine, if you will,
designing wheelchairs specifically for spinal injuries, which would be different than ones used for
amputees of those suffering from Parkinson's. By looking to the function as the requirement source,
the specifications for the system naturally come out. The current challenge to AT designers and
policy makers is how to match up functional needs to existing and proposed functional supports.
Work is being done to integrate the International Classification of Functioning, Disability and
Health from the WHO (World Health Organization, 2001) with this approach but the complications from the universe of one problem (see Section 2.7) and tying together context and tasks
make this a hard problem but one that may hold great benefits for both AT recommendations (i.e.,
matching existing AT and users needs) and design. Marcia Sheerer (Galvin and Donnell, 2002) and
RESNA/ISO standards groups (RESNA, 2015) have been working on this problem. One analogy
for the state of research is the emergence of object-oriented programming (Booch, 1991), which
holds much promise for code reuse and moving software production from a craft, always re-creating
a new wheel when a new wheel is needed, to an engineering discipline, with re-usable components
(e.g., using standard screw connectors and steel components rather than forging them).

The website of Raising the Floor, a project implementing the "AccessForAll approach, based
on achieving accessibility and digital inclusion through dynamically matching each individual's
unique needs and preferences with the resources, services, interfaces, or environments available. To
support access for all" makes this very clear:

> .....that people with disabilities are one of the most heterogeneous groups and do not fit
> neatly into diagnostic categories; that those categories can yield misleading information;
> and that the diagnostic category may only be a small factor in his or her needs and pref-
> erences. 19[19]

5.1.2   CANONICAL PAPER
Scherer, M. (2011). *Assistive Technologies and Other Supports for People With Brain Impairments.*
Springer. (Scherer, 2011)

[19]   http://raisingthefloor.org/who-we-are/our-approach/.

5.1.3   AT DESIGN EXAMPLES

I have seen the diagnosis approach often in reviewing papers for journals and conferences. Experts in computer science, new to AT for IA, sometimes build systems focused on specific causes of cognitive disabilities, resulting in requirements focused on narrow groups and using only a segment of the potential population that it could be applied to for evaluation (i.e., the autism example mentioned above). Often, when the design rationale is exposed, the work that has been done in designing to the etology of the problems to be solved does not provide insights needed to specify functionality or interface requirements for the problem solution.

5.2   I HAVE A THEORY; I HAVE A COUSIN

Short Definition: Motivating AT design is often solely based on personal relationship or solely on professional academic expertise. Relying only on either of these has consequences for broad AT abandonment and adoption.

    Longer Description: One class of designers may approach the problem of designing AT for people with cognitive disabilities by addressing the problem's space entirely theoretically and gaining understanding of the system needed through formal and academic studies of cognition and cognitive disability. The result is often elegant, intricate systems that often were abandoned or not even brought to market due to unfamiliarity with the complexity and environmental demands. One aspect of this is due to thefact that the distance between the experience of the designer and the end-user's systems are often inappropriate or ineffective in real context of use. In attempting to understand the needs of the user and the task to be performed, the system designer can take one or two naive approaches: (1) he can extrapolate from his own personal experience of someone who needs support in their day-to-day life because of cognitive disability (i.e., a child, parent, or cousin) or (2) he can establish a theory. Through daily contact the designer produces a cognitive support that perfectly fits their end-user. The problem here is that since the universe of one property characterizes the solution space of this population, the system does not generalize well and can easily fail to be adopted by others. One could label these two problems "I've got a cousin and I've got a theory."

5.2.1   THEORY

Researchers can approach this perspective as generalists, like cognitive science-based linguists, and become interested in the problems of cognitive pathological states in day-to-day life. This is, to some extent, a natural progression from the earlier work on neurology which, lacking modern imaging and sensor abilities, developed models of brain function based on pathological conditions, That is to say that hypothesizing functioning cognitive activity models are based on studying people with sensory and cognitive disabilities based on physical trauma or developmental problems. The resultant AT system for support often reflects these models and, at worst, *only* these models. This end of this axis is also tied to the problem with designing to the diagnosis rather than the functional need.

As a result, the systems produced are often difficult to configure, use and modify as the details of the model they are based on can "float" to the surface of the user interface and may require understanding the model itself to adapt or configure (Kintsch, 2002). This is apparent in some of the high-end AAC systems designed by linguists (Kintsch, 2002). So, we have one end of the Turing tar-pit (Perlis, 1982) "in which everything is possible but nothing of interest is easy."

5.2.2 COUSIN

The other end of the axis I call "I have a cousin." One of the sources for inspiration for the CLever projects was the Visions system. A special education policy lawyer in Denver, and his wife, a schoolteacher and former ARC director, had a developmentally disabled daughter, Stacie.[20] Stacie was living with the family and Bill the father was inspired (with his son) to build a system to support Stacie moving in to a condominium they purchased for her. They hired a set of programmers and produced an image and verbal prompting program to help Stacie be independent in ADLs and IADLs. This worked:

> *Stacie is now living in her own townhome with a roommate who does NOT have a disability, who does NOT provide support services for Stacie, and who pays rent to live with Stacie. As true roommates, they have become great friends and do many community activities, church activities, and social activities together. Even though Stacie is unable to read, write, tell time, understand money, or even verbalize all of her needs, she knows what to do and when to do it because the prompts remind her so that she can control her own life. She follows the step-by-step picture prompts to make dinner or go to the grocery store, or choose a recreational activity, etc. The Visions System has allowed Stacie to be a productive roommate instead of a dependent person with a disability* (from the VISIONSYSTEM.com archive1[21]).

This program was so successful for Stacie that the family decided to make a commercial system of their success (Baesman, 1999), which they sold to the IMAGINE, a Boulder Colorado nonprofit organization providing support for people with cognitive disabilities. IMAGINE (Imagine!, 2007) had, at the time, a 2,00-person residential apartment building supporting independent living. The CLever group was invited to take an investigative visit to the facility where they had purchased a Visions system, installed it in an apartment, but later abandoned it. It turned out that it was too

---

[20] I am using real names here as the Baesman family's Vision system website also used their real names. The website is gone but the internet archive allows access to old versions of the site (https://web.archive.org/web/20050308195237/http://www.thevisionssystem.com/).

[21] https://web.archive.org/web/*/www.thevisionssystem.com/ or https://web.archive.org/web/20040204134154/http://www.thevisionssystem.com/

tailored to Stacie, and not easily or inexpensively extendable for the unique needs of other people with cognitive disabilities. Too tight a fit and based on idiosyncratic desires and needs of an individual user created too much of an obstacle to broad adoption. MAPS design was, in many ways, both inspired by, and a response to, the Visions system (Figure 5.1).

Figure 5.1: Visions system logo. From Baesman and Baesman (2007).

5.2.3   CANONICAL PAPER
There really are no papers on this topic, but the best start is to start counting ratios between adopted AT and AT that is seemingly of high potential but not wide use.

5.2.4   AT EXAMPLES
There are many examples of "cousin"-based design in task support besides Visions. The problem is, I think, based on both a lack of research-based understanding of the problem and too tight a focus on a single user (a seemingly appropriate response to the universe of one problem)

The examples of "theory"-based adoption/abandonment problem prone systems are often centered in AAC domain. On a tour of special education AT devices I was told stories about abandonment of complex tools to support speech (as the very complex one used by the physicist Steven Hawking, and that exact model was on a shelf with abandoned, overly complex devices), resulting in several very expensive systems in a corner of a AT storage room, abandoned due to the theoretical foundations of the system "floating" to the interface (Kintsch, 2002).

5.2.5   CONCLUSIONS
What to do? If your primary motivation and background is in academia and R&D (especially focused on the technology and not the domain), then the best thing is to spend time with people that are the intended users of your system. More than that, seek out end-user groups and end-user caregiver support groups; seek out other specialists in stakeholders, such as rehabilitation professionals, family and special education application professionals. Don't just interview them, spend time with them and look at what they are doing to accomplish what you intend to replace.

If you're coming to AT design from a more personally motivated perspective, your child, co-worker, friend, and yes, your cousin, may be the model for the system you make. You will, if successful, end up with a very nicely tailored system that does exactly what your end-user needs. This may be a good place to start. The next step would be to talk to professionals in the field and talk about what is needed to make this functionally workable for others. Better still is to think about this at design time, but self-motivated system creators don't have this as the primary motivation to continue, often with no intent to recoup time and money spent on it beyond the satisfaction of helping another.

5.3   ISLANDS OF ABILITY

Short Definition: High-functioning AT end-users often have: (1) personally heterogeneous cognitive and physical disabilities; and (2) abilities that may widely vary over time.

Longer Description: The "universe of one" conceptualization includes the empirical finding that (1) *unexpected islands of abilities* exist: clients can have unexpected skills and abilities that can be leveraged to ensure a better possibility of task accomplishment; and (2) *unexpected deficits of abilities* exist (Figure 5.2). These anomalous deficits and abilities can be static, as in tasks that are always difficult to accomplish; or dynamic, as in stress potentiated, fatigue triggered, or even seasonal.

In the MAPS system evaluation I was observing the use of the system by a young lady with intellectual disabilities that were sufficient to bar her from employment except in simple tasks. More than that she required, and was given, a job coach to help her learn to sort used clothes on a rack at a local used household and clothing store (see Figure 2.4). The coach expected, using traditional rehabilitation teaching techniques, for it to take several months for her to master the sequences of actions required; with the MAPS system this was halved. More to the point, one day I was watching her use the system and encountered her grandmother and spent a short time talking to her about her granddaughter and passing the time (actually I was gathering family history and probing for the impact of the system in the family structure). The young lady came from a quite wealthy family and in their large home in the mountains the father has installed a home theater, something that in 2004 was quite rare (and technically complex) . The grandmother told me she was having problems starting up and setting the volume and lighting in the room, when the young lady came in and explained to the grandmother how to do it while doing it for her. This the grandmother found quite remarkable. To some extent this was a generational difference—remote controls are to millennials as dial telephones were to the greatest generation. But beyond this, her facility with the controls was remarkable. This particular young lady had an operatic level singing voice, and I think one of the insights that can be gained from my work with her is that the term "idiot savant" is widely inappropriately used. The dimensions of ability that we all have differ in so many domains.

Islands of abilities in seas of deficits:
Unexpected abilities that can be leveraged

Islands of deficits in seas of abilities:
causes of unexpected activity failures

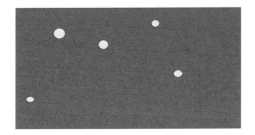

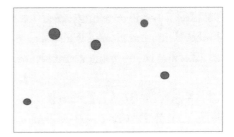

Figure 5.2: Islands of need and ability. From Cole (2006).

Experts in this domain, such as Elliot Cole, did not find this very remarkable. To have intellectual disabilities, or cognitive disabilities, is not a matter of turning a single control on cognitive abilities down. There are many facets of ability and needs that differ from person to person, and especially in people with cognitive disabilities over time. This time may be a day, a week, context dependent, and even developmental (we all grow in so many dimensions over our lifetime).

These islands, static and dynamic, appear in many narratives of people with cognitive disabilities and disabilities in general (Scherer, 1996). Having a hard day, or having a hard week, is not solely the domain of typically abled people, but the impact and ability to compensate may be diminished for those with various disabilities.

An example is the MAPS end user who held an outside non-sheltered job, took the bus to and from the job, shopped, and even had a bank account that he managed. He did his own laundry but could never (he was in his 30's) learn how to fold and put away his own clothes. Hanging pants was particularly challenging to him. This was worked around by creating specific scripts that changed as he became slowly more proficient, however pants hanging never was collapsed in the script, so for the experimental period he never mastered unaided hanging pants on hangers: Islands of inability.

The island model sometimes works both ways. I have a friend who has been paraplegic since his early 20's, and in the 1980's we were part of a group of sailing enthusiasts. Because he was always quite athletic his upper body strength was remarkable, and ten of us had rented a 50' sailing boat to go from southern Maine to Nova Scotia. We came into some rough weather on the second day of sailing and many of us went below decks while those on watch stayed in the cockpit or on deck. While below the winds and water caused the boat to move abruptly back and forth, up and down. To move from aft to forward was a bumpy zig-zag, but not for my friend—he grabbed the railing in the ceiling and moved easily and gracefully—helping me prepare the noon meal. That day we all asked him for help: Islands of ability.

Compounding the problem is that this variability may be a contextually driven. Driving while using a phone or texting are an excellent and dangerous example. Having a broken leg or arm is similar. Being intoxicated is also a case where my intellectual capacity is diminished but will not be so 12 hours later—think about airline pilots and flight rules.

5.3.1 CANONICAL PAPER

Cole, E. (2013). *Patient-Centered Design of Cognitive Assistive Technology for Traumatic Brain Injury Telerehabilitation. Synthesis Lectures on Assistive, Rehabilitative.* (Cole, 2013).

5.3.2 AT DESIGN EXAMPLES

This is an item where examples of existing AT can be provided, as much as another cautionary tale about (1) knowing deeply the target population, stakeholders, and tasks, and (2) the importance of flexibility in the system.

Designing for the minimum applicable amount of functionality may lead to boredom and abandonment; designing too high can lead to frustration and abandonment. The earlier sections of this book presented tools like personalization, design for failure, and intelligence augmentation as ways not to fall into this trap, but the most important tool is to spend time with the end-user—with lots of end users and caregivers, and AT experts.

PART 3

Further Study

The previous section gave, for each perspective or framework, enough information to begin implementing these ideas in your design work. However, each one is, to a greater or lesser degree, imbedded in a matrix of research and use. The following section lists a collection of articles representing research that, in some cases, supported the generation of the topics discussed, and in others extended or implemented the concepts.

CHAPTER 6

Source Documents for Each Idea

6.1 ARTIFICIAL INTELLIGENCE (AI) INTELLIGENCE AUGMENTATION (IA)

Doug Englebart is the creator and evangelist for so much of modern computer interfaces. Mice, drag and drop, the desktop, all these were part of implementing his vision of computers as helpmates. The key papers are listed below. Beyond this there is a very interesting video of his early work "the mother of all demos" which you can see online.[22]

> Engelbart, D.C. (1962). *Augmenting Human Intellect: A Conceptual Framework*.

> Engelbart, D. C. and W. K. English (1968). A research center for augmenting human intellect. *Proceedings of the AFIPS Fall Joint Computer Conference*. Washington, DC; The Thompson Book Company, pp. 395–410.

> Engelbart, D. C. (1988). A conceptual framework for the augmentation of man's intellect. In *Computer-Supported Cooperative Work: A Book of Readings*. I. Greif. San Mateo, CA:Morgan Kaufmann Publishers, pp. 35–66.

> Engelbart, D. C. (1990). Knowledge-domain interoperability and an open hyper-document system. *Proceedings of ACM CSCW'90 Conference on Computer-Supported Cooperative Work*. F. Halasz, Ed. New York:ACM, pp. 143–156.

> Engelbart, D. C. (1995). "Toward augmenting the human intellect and boosting our collective IQ." *Communications of the ACM* 38(8): 30-33.

> Engelbart, D. and K. Hooper (1988). The augmentation system framework. In *Interactive Multimedia*. S. Ambron and K. Hooper, Eds. Redmond, WA:Microsoft Press, pp. 15–31.

Greg Vanderheiden is a pioneer in assistive technology and now leads the Global Public Inclusive Infrastructure project[23] and Cloud4All[24] framework. This article was the basis for much work done in the spirit of this book.

---

[22] http://tinyurl.com/kue79ck.
[23] http://gpii.net/.
[24] http://www.cloud4all.info/.

Vanderheiden, G. (1992). A brief look at technology and mental retardation in the 21st century. In *Mental Retardation in the Year 2000*. L. Rowitz (Ed.). New York:Springer-Verlag, Inc., pp. 268–278.

Licklider, J. C. R. (1960). "Man-computer symbiosis." *IRE Transactions on Human Factors in Electronics HFE-1*, pp. 4–11.

Ashby, W. R. (1956). "An Introduction to Cybernetics." London:Chapman and Hall 16: 295.

Markoff, J. (2005). *What the Dormouse Said: How the Sixties Counterculture Shaped the Personal Computer Industry*. New York:Penguin Books.

6.2 DESIGN FOR FAILURE

Patterson, D. A., G. Gibson, and R. H. Katz (1988). "A case for redundant arrays of inexpensive disks (RAID)." *ACM SIGMOD Record* 17: 109–116.

Sommerville, I. (2001). *Software Engineering*. Reading, MA:Addison-Wesley.

Lewis, C. H. and D. A. Norman (1986). Designing for error. In *User Centered System Design, New Perspectives on Human-Computer Interaction*. D. A. Norman and S. W. Draper, Eds. Hillsdale, NJ:Lawrence Erlbaum Associates, pp. 411–432.

This is an excellent general study of failure and systems:

Petroski, H. (1985). *To Engineer Is Human: The Role of Failure in Successful Design*. New York:St. Martin's Press.

6.3 DISTRIBUTED COGNITION

Salomon, G., Ed. (1993). *Distributed Cognitions: Psychological and Educational Considerations*. Cambridge, UK, Cambridge University Press.

Rogers, Y. (1997) "A brief introduction to distributed cognition." Discussion paper, Interact Lab, School of Cognitive and Computing Sciences, University of Sussex.

Perry, M. (2003). Distributed cognition. In *HCI Models, Theories and Frameworks, Toward a Multidisciplinary Science*. J. M. Carroll, Ed. Burlington, VT:Morgan Kaufman, pp. 193–223.

Perkins, D. N. (1993). Person-plus: A distributed view of thinking and learning. *Distributed Cognitions Psychological and Educational Conciderations*. G. Solomon, Ed. Cambridge,UK:Cambridge University Press, pp. 88–110.

Pea, R. (1993). Practices of distributed intelligence and designs for education. In *Distributed Cognitions: Psychological and Educational Considerations*. G. Salomon, Ed. Cambridge, UK:Cambridge University Press, pp. 47–87.

Nickerson, R. S. (1993). On the distribution of cognition: some reflections. In *Distributed Cognitions Psychological and Educational Considerations*. G. Salomon, Ed. Cambridge, UK:Cambridge University Press, pp. 229-260.

Hollan, J., E. Hutchins, and D. Kirsch (2001). Distributed cognition: Toward a new foundation for human-computer interaction research. In *Human-Computer Interaction in the New Millennium*. J. M. Carroll, Ed. New York:ACM Press, pp. 75–94.

Fischer, G. and S. Konomi (2005). Innovative media in support of distributed intelligence and lifelong learning. *Proceedings of the Third IEEE International Workshop on Wireless and Mobile Technologies in Education*. Tokushima, Japan, IEEE Computer Society, pp. 3-10.

Fischer, G., E. Arias, S. Carmien, H. Eden, A. Gorman, S. i. Konomi, and J. Sullivan. (2004). "Supporting collaboration and distributed cognition in context-aware pervasive computing environments (Paper presented at the 2004 *Meeting of the Human Computer Interaction Consortium "Computing Off The Desktop"*)." Winter Park, CO

Ernest Hutchins' book on distributed cognition in naval vessels coastal navigation is the best detailed introduction to how DC actually works.

Hutchins, E. (1996). *Cognition in the Wild*. Cambridge MA:MIT Press.

Carmien, S., I. Kollar, G. Fischer, and F. Fischer (2007). The interplay of internal and external scripts — a distributed cognition perspective. In *Scripting Computer-Supported Learning—Cognitive, Computational, and Educational Perspectives*. F. Fischer, I. Kollar, H. Mandl and J. M. Haake, Ed. New York:Springer, pp. 303–326.

Carmien, S. (2007). *Leveraging Skills into Independent Living- Distributed Cognition and Cognitive Disability*. Saarbrücken, VDM Verlag Dr. Mueller e.K.

Rogers, Y. and J. Ellis (1994). "Distributed cognition: An alternative framework for analysing and explaining collaborative working." *Journal of Information Technology* 9, pp. 119–128.

Nardi, B. (1996). *Context and Consciousness and Consciousness Activity and Human-Computer Interactions*. Cambridge, MA:The MIT Press.

Neressian, N. J., Newstetter, W. C., Kurz-Milcke, E., and Davies, J. (2003). A mixed-method approach to studying distributed cognition in evolving environments. *The fifth International Conference on Learning Sciences*. Seattle, WA.

6.4   SCAFFOLDING

Knops, H. and C. Bühler (1999). *Assistive Technology on the Threshold of the New Millennium*. Amsterdam, Washington, DC, Tokyo:IOS Press

Carmien, S. and R. Koene (2009). Distributed intelligence and scaffolding in support of cognitive health. *13th International Conference on Human-Computer Interaction (HCII 09)* in the parallel session "Cognitive Accessibility and Cognitive Support" in the *5th International Conference on Universal Access in Human-Computer Interaction (UAHCI)*. San Diego, CA, Springer. LNCS vol. 5614, pp. 334–343.

Rosson, M. B., J. M. Carroll, and R. K. E. Bellamy (1990). Smalltalk Scaffolding: A case study of minimalist instruction. *CHI '90*. J. C. Chew and J. Whiteside, Eds. New York:ACM, pp. 423–429.

Guzdial, M. (1994). "Software-realized scaffolding to facilitate programming for science learning." *Interactive Learning Environments* 4:1 pp. 001-044. DOI: 10.1080/1049482940040101.

Davies, E. A. and N. Miyake (2004). "Special issue: Scaffolding." *The Journal of the Learning Sciences* 13(3): 265–451.

O'Neill, B., K. Moran, and A. Gillespie (2010). Scaffolding rehabilitation behavior using a voice-mediated assistive technology for cognition. *Neuropsychological Rehabilitation*. Vol. 20, pp. 509–527.

Wood, D. and D. Middleton (1975). "A study of assisted problem-solving." *British Journal of Psychology* 66: 181–191.

Wood, D., J. S. Bruner, and G. Ross (1976). "The role of tutoring in problem-solving." *Journal of Child Psychology and Psychiatry* 17: 89–100.

Mohamad, Y. (2005). Integration von Emotionaler Intelligenz in Interface-Agenten am Beispiel einer Trainingssoftware für lernbehinderte Kinder. Ph.D., Rheinisch-Westfälischen Technischen Hochschule Aachen.

Mohamad, Y., C. A. Velasco, S. Damm, and H. Tebarth (2004). "Cognitive training with animated pedagogical agents (TAPA) in children with learning disabilities." *Computers Helping People with Special Needs*. 9th International Conference, ICCHP 2004. Paris, France, July 7-9, 2004 629.

6.5   SITUATED ACTION/COGNITION

Suchman, L. (1987). *Plans and Situated Actions: The Problem of Human-Machine Communication*. Cambridge, UK:Cambridge University Press.

Suchman, L. A. (2007). *Human-Machine Recongfigurations, Plans and Situated Actions*, 2nd edition. New York:Cambridge University Press.

6.6   SOCIO-TECHNICAL ENVIRONMENTS

Scacchi, W. (2004). Socio-technical design. *The Encyclopedia of Human-Computer Interaction*. W. S. Bainbrigde, Berkshire Publishing Group, from https://www.interaction-design.org/literature/book/the-encyclopedia-of-human-computer-interaction-2nd-ed.

Mumford, E. (1987). Sociotechnical systems design: Evolving theory and practice. In *Computers and Democracy*. G. Bjerknes, P. Ehn, and M. Kyng, Eds. Aldershot, UK, Avebury, pp. 59–76.

Mumford, E. (2000). Socio-technical design: An unfulfilled promise or a future opportunity. *Proceedings of the IFIP TC9 WG9.3 International Conference on Home Oriented Informatics and Telematics*, "IF at Home: Virtual Influences on Everyday Life": Information, Technology and Society, June 2000.

Mumford, E. (2000). A socio-technical approach to systems design. *Requirements Engineering* 5: 59–77.

Erickson, T., D. N. Smith, W. A. Kellogg, M. Laff, and E. Brander (1999). A socio-technical approach to design: Social proxies, persistent conversations, and the design of babble. *Proceedings of ACM CHI 99 Conference on Human Factors in Computing Systems (CHI '99)*. Pittsburgh, PA

Carmien, S. (2011). Socio-technical environments and assistive technology. In *Socio-technical Networks: Science and Engineering Design*. F. Hu, A. Mostashari, J. Xie, Eds. Boca Raton, FL:Taylor and Francis LLC, CRC Press, pp. 167–180.

6.7   UNIVERSE OF ONE

Jakovljević, M. and L. Ostojić (2013). "Comorbidity and multimorbidity in medicine today: challenges and opportunities for bringing separated branches of medicine closer to each other." *Psychiatria Danubina* 25 (Suppl 1): 18–28.

Mc Sharry, J. (2014). "Challenges in managing multiple conditions: The patient experience of multimorbidity." *The European Health Psychologist* 16(6): 224–227.

Erikson, E. (1958). *The Nature of Clinical Evidence. Evidence and Inference.* Glencoe, IL:Free Press of Glencoe.

6.8   WICKED PROBLEMS

Rittel, H. and M. M. Webber (1984). Planning problems are wicked problems. In *Developments in Design Methodology*. N. Cross, Ed. New York:John Wiley & Sons, pp. 135–144.

Rittel, H. (1984). Second-generation design methods. In *Developments in Design Methodology*. N. Cross, Ed. New York:John Wiley & Sons, pp. 317–327.

6.9   DYADS

Carmien, S. and A. Kintsch (2006). Dual user interface design as key to adoption for computationally complex assistive technology. *RESNA Annual Conference*, Atlanta GA.

6.10   IMPORTANCE OF REPRESENTATION

Simon, H. A. (1996). *The Sciences of the Artificial*. Cambridge, MA:The MIT Press.

Carmien, S. and E. Wohldmann (2008). "Mapping images to objects by young adults with cognitive disabilities." *Research in Developmental Disabilities* 29: 149–157.

6.11   TOOLS FOR LIVING AND TOOLS FOR LEARNING

Carmien, S. and G. Fischer (2005). Tools for living and tools for learning. *Proceedings of the HCI International Conference (HCII)*, Las Vegas, July 2005 (CD).

6.12   PLANS AND ACTIONS

Schank, R. C. and R. P. Abelson (1977). *Scripts, Plans, Goals, and Understanding*. Hillsdale, NJ:Lawrence Erlbaum Associates, Inc.

Reason, J. (1990). *Human Error*. Cambridge, UK:Cambridge University Press.

Norman, D. A. (1983). "Design rules based on analyzes of human error." *Communications of the ACM* 26(4): 254-258.

Lewis, C. H. and D. A. Norman (1986). Designing for error. In *User Centered System Design, New Perspectives on Human-Computer Interaction*. D. A. Norman and S. W. Draper, Eds. Hillsdale, NJ:Lawrence Erlbaum Associates, pp. 411–432.

Dekker, S. (2006). *The Field Guide to Undersanding Human Error*. Burlington, VT:Ashgate.

6.13   LOW-HANGING FRUIT

Juran, J. M. (1950). "Pareto. Lorenz, Cournot Bernoulli, Juran and others." *Industrial Quality Control* 2(October 1950): 25.

6.14   METADESIGN

Maturana, H. R. (1997). "Metadesign." From http://www.hum.auc.dk/~rasand/Artikler/metadesign.htm.

Maturana, H. R. and F. J. Varela (1987). *The Tree of Knowledge: The Biological Roots of Human Understanding*. Boston, MA:Shambhala.

Giaccardi, E. and G. Fischer (2008). "Creativity and evolution: A metadesign perspective." *Digital Creativity* 19(1): 19–32.

Giaccardi, E. (2005). "Metadesign as an emergent design culture." *Leonardo* 38(4): 342–349.

Fischer, G., Giaccardi, E., and D. Fogli. (2005). "The concept of seed in meta-design: Nature, processes, and evolution." The Center for Lifelong Learning and Design, University of Colorado.

Fischer, G. and E. Scharff (2000). Meta-design—Design for designers. *3rd International Conference on Designing Interactive Systems (DIS 2000)*, New York, ACM.

Fischer, G. and T. Herrmann (2008). "Sociotechnical systems: A meta-design perspective." *International Journal of Sociotechnology and Knowledge Development (IJSKD)* 3(1): 1–33.

Fischer, G. and E. Giaccardi (2006). Meta-design: A framework for the future of end user development. In *End User Develop—End User Development — Empowering people to flexibly employ advanced information and communication technology*. H. Lieberman, F. Paternò and V. Wulf, Eds. Dordrecht, The Netherlands:Kluwer Academic Publishers, pp. 427–457.

6.15 PERSONALIZATION

Lieberman, H., F. Paterno, and V. Wulf, Eds. (2006). *End User Development*. Dordrecht, The Netherlands:Kluwer Publishers.

Fisk, A. D., Rogers, W. A., Charness, N., Czaja, S. J., and Sharit, J. (2009). *Designing for Older Adults: Principles and Creative Human Factors Approaches*, 2nd edition, CRC Press, Taylor and Francis Croup. Boca Raton, FL.

Apple, i. (2009). Apple human interface guidelines. Cupertino, CA, Apple.

6.16 SYMMETRY OF IGNORANCE

Fischer, G. (2000). "Social creativity, symmetry of ignorance and meta-design." *Knowledge-Based Systems Journal* (Special Issue on Creativity & Cognition), 13(7-8): 527–537.

Fischer, G., P. Ehn, Y. Engeström, and J. Virkkunen (2002). Symmetry of ignorance and informed participation. *Proceedings of the Participatory Design Conference (PDC'02)*, Malmö University, Sweden, CPSR.

Rittel, H. W. J. (1972). "On the planning crisis: Systems analysis of the first and second generations." *Bedrifts Økonomen* (Norway), 8(107).

Buxton, W. (2002). Less is more (more or less). In *The Invisible Future—The Seamless Integration of Technology in Everyday Life*. P. J. Denning (ed.). New York, McGraw-Hill: 145–179.

6.17   DIAGNOSIS AND FUNCTIONALITY

Scherer, M. (2011). *Assistive Technologies and Other Supports for People With Brain Impairment*, New York:Springer Publishing Company.

Scherer, M. J. (1996). *Living in the State of Stuck: How Technology Impacts the Lives of People with Disabilities*. Cambridge:Brookline Books.

Scherer, M. J. and Galvin, J.C. (1994). "Matching people with technology." *Rehab Management* 7(2): 128–130.

Scherer, M. J. and J. C. Galvin. (Editor) (1996). *Evaluating, Selecting, and Using Appropriate Assistive Technology*. Gaithersburgh, MD:Aspen Publishers.

World Health Organization (2001). International Classification of Functioning, Disability and Health (ICF), World Health Organization (WHO). Deneva, Switzerland.

6.18   I HAVE A THEORY; I HAVE A COUSIN

You could be the first! Maybe we could write it together.

6.19   ISLANDS OF ABILITY

Cole, E. (2013). *Patient-Centered Design of Cognitive Assistive Technology for Traumatic Brain Injury Telerehabilitation*. San Rafael, CA:Morgan & Claypool.

CHAPTER 7

Conclusion

What I have tried to do in this book is to articulate and provide tools to the designer of assistive technology, and other appropriate design domains, so that you can implement and expand them. One interesting thing about this set is the heterogeneity of their provenance. As a result, some of the source texts come from divergent viewpoints and those new to them may find them difficult to digest. To mitigate this I have tried, in several areas, to give detailed examples of how they may be used.

Stretching is good, as an example both Meta Design and Distributed Cognition were hard for me to initially digest, but the effort has enriched my "toolkit," and this motivated me to write this book. Creating applications for people with disabilities is not a trivial or obvious task; there is a lot to learn, and I hope that this book and the others in the series provide useful guidance and spark for you in your work.

Bibliography

AbleLink (2002). "Visual impact memory and skill prompter." From http://www.ablelinktech.com/ProductPage.asp?SelectedProduct=VisualImpact. 81, 110

AbleLink (2007). "Ablelink Technologies website." Retrieved 2003, from http://www.ablelinktech.com. 110

ADA (August 2015). "ADA checklist for existing facilities." From http://www.adachecklist.org/doc/fullchecklist/ada-checklist.pdf. 3

ADA. (1990). "American with Disabilities Act of 1990." From http://www.usdoj.gov/crt/ada/ada-hom1.htm. 3

Alexander, C., S. Ishikawa, M. Silverstein, M. Jacobson, I. Fiksdahl-King, and S. Angel (1977). *A Pattern Language: Towns, Buildings, Construction*. New York:Oxford University Press. 10

Andersson, E. (2012). Motion classification and step length estimation for GPS/INS pedestrian navigation. EES Examensarbete/Master Thesis. 37

Apple, i (2009). *Apple Human Interface Guidelines*. Cupertino, CA:Apple. 112

Ashby, W. R. (1956). "An introduction to cybernetics." New York:J. Wiley, 16: 295. DOI: 10.5962/bhl.title.5851.

Asimov, I. (1959). *Nine Tomorrows*, Greenwich, CT:Fawcett Crest. 90

ASSISTANT (2012a). "ASSISTANT Website." Retrieved January 2013, from http://www.aal-assistant.eu/. 25, 28, 106

ASSISTANT (2012b). Description of Work (DOW). http://www.aal-assistant.eu/.

ASSISTANT Consortium (2015). D 3.3.4 Final user evaluation report AAL CMU. http://www.aal-assistant.eu/. 107

Assistant Project (2015). **A**iding **SuS**tainable **I**ndependent **S**enior **TrA**vellers to **N**avigate in **T**owns. EU, AAL. 4, 28, 30, 31, 33, 36, 39, 78, 112, 128

Baesman, B. (1999). Interview, the inventor of the Visions system. 134

Baesman, B. (2000). "The visions system." From http://www.thevisionssystem.com. 91

Baesman, B. and N. Baesman. (2007). "Visions System website." From http://www.thevisionssystem.com. 135

Bateson, G. (1972). *Steps to an Ecology of Mind.* Chicago, IL:University of Chicago. 44

Bernard, H. R. (2000). *Social Research Methods.* Thousand Oaks, California, Sage Publications, Inc. 61

Bernard, H. R. (2002). *Research Methods in Anthropology.* Walnut Creek, CA:Altamira Press. 61

Beukelman, D. and P. Mirenda (1998). *Augmentative and Alternative Communication.* Baltimore, MD:Paule H. Brookes Publishing. 81, 82

Bodine, C. (2003). Personal communication. Assistive Technology Partners. S. Carmien. Denver. 5

Booch, G. (1991). *Object Oriented Design with Applications.* Redwood City, CA:The Benjamin Cummings Publishing Company, Inc. 132

Braddock, D. (2006). 2006 Cognitive Disability in U.S., University of Colorado, Coleman Institute. 105

Braun, J. (2003). "Natural scenes upset the visual applecart," *TRENDS in Cognitive Sciences* 7(1): 7–9. DOI: 10.1016/S1364-6613(02)00008-6. 85

Bronnec, E. (2015). Ofvelopptment des Ventes Fruiu et Légumes, Carrefour. 17

Buxton, W. (2002). Less is more (more or less). In *The Invisible Future—The Seamless Integration of Technology in Everyday Life.* P. J. Denning (ed.). New York:McGraw-Hill: 145–179. 126

Carmien, S. (2002). "Express Reservations website." 2015. From http://tinyurl.com/gpnk3b3. 126

Carmien, S. (2004a). Doctoral consortium: MAPS: creating socio-technical environments in support of distributed cognition for people with cognitive impairments and their caregivers. Vienna, Austria:ACM press. 27

Carmien, S. (2004b). "MAPS Website." From http://www.cs.colorado.edu/~l3d/clever/projects/maps.html. 25, 76

Carmien, S. (2004c). MAPS: creating socio-technical environments in support of distributed cognition for people with cognitive impairments and their caregivers. *Extended Abstracts of the 2004 Conference on Human Factors and Computing Systems.* Vienna, Austria:ACM Press, pp. 1051–1052. DOI: 10.1145/985921.985974.

Carmien, S. (2005). MAPS/Lifeline Project. A. Gorman. University of Colorado. 35, 87

Carmien, S. (2006a). Assistive technology for people with cognitive disabilities—Artifacts of distributed cognition. *CHI06 Workship of Desinging Technology for People with Cognitive Impariments at the CHI'06 Conference on Human Factors in Computing Systems,* Montreal, PQ.

Carmien, S. (2006b). Socio-technical environments supporting distributed cognition for people with cognitive disabilities. Ph.D. dissertation, University of Colorado at Boulder. 18, 21, 50, 76, 86, 93, 98, 109

Carmien, S. (2007). *Leveraging Skills into Independent Living- Distributed Cognition and Cognitive Disability*. Saarbrücken:VDM Verlag Dr. Mueller e.K. 19, 23, 24, 75, 114

Carmien, S. (2009). Technical annex for type A or B projects; Project title: Financial decision making aid for elders (FIDEMAID). Convocatoria de ayudas de Proyectos de Investigación Fundamental no orientada. San Sebastian, Spain, Tecnalia. 76, 77

Carmien, S. (2011). Socio-technical environments and assistive technology. Socio-technical networks: Science and engineering design. F. Hu, Mostashari, A., Xie, J. . Boca Raton, FL:Taylor and Francis LLC, CRC Press, pp. 167–180. 22, 23, 75

Carmien, S. and A. M. Cantara (2012). Diagnostic and accessibility based user modeling. In *User Modeling and Adaptation for Daily Routines Providing Assistance to People with Special Needs*. E. MartÌn, P.A Haya, R.M. Carro, Eds. Springer, p. 232. 117, 118, 123

Carmien, S., M. Dawe, G. Fischer, A. Gorman, A. Kintsch, and J. F. Sullivan (2005). "Socio-technical environments supporting people with cognitive disabilities using public transportation." *Transactions on Human-Computer Interaction (ToCHI)* 12(2): 233-262. DOI: 10.1145/1067860.1067865. 4, 23

Carmien, S., R. DePaula, A. Gorman, and A. Kintsch (2004). "Increasing workplace independence for people with cognitive disabilities by leveraging distributed cognition among caregivers and clients." *Computer Supported Cooperative Work (CSCW)* 13(5–6): 443–470. 93

Carmien, S. and G. Fischer (2005). Tools for living and tools for learning. *Proceedings of the HCI International Conference (HCII)*, Las Vegas, July 2005 (CD). 86, 91

Carmien, S. and G. Fischer, (2008). Design, adoption, and assessment of a socio-technical environment supporting independence for oersons with cognitive disabilities. *ACM Conference on Computer-Human Interaction CHI08*, Florence, Italy. 20, 21

Carmien, S. and A. Garzo (2014). Elders using smartphones: a set of research based heuristic guidelines for designers. *16th International Conference on Human-Computer Interaction (HCII 09)* in the parallel session "Cognitive Accessibility and Cognitive Support" in the *8th International Conference on Universal Access in Human-Computer Interaction (UAHCI)*. Heraklion, Crete, Greece:Springer LNCS, Lecture Notes in Computer Science 8514: 26–37. 128

Carmien, S. and A. Gorman (2003). Creating distributed support systems to enhance the quality of life for people with cognitive disabilities. *2nd International Workshop on Ubiquitous Computing for Pervasive Healthcare Applications (UbiHealth 2003)*, Seattle, WA. 34

Carmien, S. and A. Kintsch (2006). Dual user interface design as key to adoption for computationally complex assistive technology. *RESNA Annual Conference*, Atlanta GA. 76

Carmien, S. and R. Koene (2009). Distributed intelligence and scaffolding in support of cognitive health. *13th International Conference on Human-Computer Interaction (HCII 09)* in the parallel session "Cognitive Accessibility and Cognitive Support" in the *5th International Conference on Universal Access in Human-Computer Interaction (UAHCI)*. San Diego, CA:Springer. LNCS vol. 5614: 334–343.

Carmien, S., I. Kollar, G. Fischer, and F. Fischer (2007). The interplay of internal and external scripts — a distributed cognition perspective. *Scripting Computer-Supported Learning— Cognitive, Computational, and Educational Perspectives*. F. Fischer, I. Kollar, H. Mandl and J. M. Haake, Eds. New York:Springer, pp. 303–326. 50, 87

Carmien, S. and E. Wohldmann (2008). "Mapping images to objects by young adults with cognitive disabilities." *Research in Developmental Disabilities* 29: 149–157. DOI: 10.1016/j.ridd.2007.02.003. 81, 83

Carroll, J. M. (2003). *HCI Models, Theories and Frameworks, Toward a Multidisciplinary Science*, San Francisco, CA:Morgan Kaufmann Publishers. 45

Center for Universal Design (2011). "The 7 Principles of Universal Design." 69

CLever (2004). "CLever: Cognitive Levers—Helping people help themselves." From http://www.cs.colorado.edu/~l3d/clever. 7, 9, 23, 27, 87

Coakes, E., D. Willis, and R. Lloyd-Jones (1999). *The New SocioTech: Graffiti on the Long Wall*. London, Springer Verlag. 59

Cockburn, A. (2001). *Agile Software Development*. Reading, MA:Addison-Wesley. 73

Codd, E. F. (1990). *The Relational Model for Database Management* (Version 2 ed.), Addison Wesley Publishing Company. 111

Cole, E. (1997). "Cognitive prosthetics: an overview to a method of treatment." *NeuroRehabilitation* 12(1): 31–51. 5, 23

Cole, E. (2006). Patient-centered design as a research strategy for cognitive prosthetics: Lessons learned from working with patients and clinicians for 2 decades. *CHI 2006 Workshop on Designing Technology for People with Cognitive Impairments*, Montreal, Canada. 68, 70, 137

Cole, E. (2013). *Patient-Centered Design of Cognitive Assistive Technology for Traumatic Brain Injury Telerehabilitation.* Morgan & Claypool. DOI: 10.2200/S00478ED-1V01Y201302ARH003. 3, 25, 138

Coleman. (2004). "Coleman Institute for Cognitive Disabilities website." from http://www.cu.edu/ColemanInstitute. 8

Csikszentmihalyi, M. (1990). *Flow: The Psychology of Optimal Experience.* New York:HarperCollins Publishers. 51, 52

Czarnuch, S., S. Cohen, V. Parameswaran, and A. Mihailidis (2013). "A real-world deployment of the COACH prompting system." *Journal of Ambient Intelligence and Smart Environments* 5: 463–478. 42

Czarnuch, S. and A. Mihailidis (2012). "The COACH: A real-world efficacy study." *Alzheimer's & Dementia* 8: P446. DOI: 10.1016/j.jalz.2012.05.1188. 42

Davies, D., S. Stock, and M. L. Wehmeyer (2002). "Enhancing independent task performance for individuals with mental retardation through use of a handheld self-directed visual and audio prompting system." *Education and Training in Mental Retardation and Developmental Disabilities* 37(2): 209–218. 82

Davies, E. A. and N. Miyake (2004). "Special issue: Scaffolding." *The Journal of the Learning Sciences* 13(3): 265–451.

Dekker, S. (2006). *The Field Guide to Undersanding Human Error.* Burlington, VT:Ashgate. 26, 39

Deming, W. E. (1950). "Speach at Mt. Hakone Conference Center in August 1950." From http://www.jsdstat.com/Statblog/wp-includes/Hakone.pdf. 103

DePaula, R. (2002). "Web2gether Website." Retrieved March, 2002, from web2gether.cs.colorado.edu:7001/. 73

DePaula, R. (2004). The construction of usefulness: How users and context create meaning with a social networking system. Ph.D. Dissertation, University of Colorado at Boulder. 62

Doerry, E. (1995). Evaluating distributed environments based on communicative efficacy. *Conference Companion on Human Factors in Computing Systems.* Denver, CO:ACM: 47–48. 57

Engelbart, D. (1962). Augmenting human intellect: A conceptual framework. *SRI Summary Report AFOSR-3223.* Air Force Office of Scientific Research, Washington DC. 18, 27

Engelbart, D. and K. Hooper (1988). The Augmentation System Framework. *Interactive Multimedia.* S. Ambron and K. Hooper. Redmond, WA:Microsoft Press, pp. 15–31.

Engelbart, D. C. (1988). A Conceptual Framework for the Augmentation of Man's Intellect. *Computer-Supported Cooperative Work: A Book of Readings*. I. Greif. San Mateo, CA:Morgan Kaufmann Publishers, pp. 35–66.

Engelbart, D. C. (1990). Knowledge-Domain Interoperability and an Open Hyperdocument System. *Proceedings of ACM CSCW'90 Conference on Computer-Supported Cooperative Work*. F. Halasz. New York, ACM, pp. 143–156. DOI: 10.1145/99332.99351.

Engelbart, D. C. (1995). "Toward augmenting the human intellect and boosting our collective IQ." *Communications of the ACM* 38(8): 30-33. DOI: 10.1145/208344.208352. 87

Engelbart, D. C. and W. K. English (1968). A research center for augmenting human intellect. *Proceedings of the AFIPS Fall Joint Computer Conference*. Washington, DC:The Thompson Book Company: 395–410. DOI: 10.1145/1476589.1476645.

Erickson, T., D. N. Smith, W. A. Kellogg, M. Laff, and E. Brander (1999). A sociotechnical approach to design: Social proxies, persistent conversations, and the design of babble. *Proceedings of ACM CHI 99 Conference on Human Factors in Computing Systems (CHI '99)*.

Ericsson, A. K. (2003). "Exceptional memorizers: made, not born." *Trends in Cognitive Science* 7(6): 233–235. DOI: 10.1016/S1364-6613(03)00103-7. 81

Erikson, E. (1958). *The Nature of Clinical Evidence. Evidence and Inference.* Glencoe, IL:Free Press of Glencoe. 25

EU4ALL (2007). "European unified approch for assisted longlife learning." From http://www.eu4all-project.eu/www/—overview. 112

Fischer, G. (1998). "Seeding, evolutionary growth and reseeding: Constructing, capturing and evolving knowledge in domain-oriented design environments." *Automated Software Engineering* 5(4): 447–464. DOI: 10.1023/A:1008657429810. 73

Fischer, G. (1999). "A group has no head — Conceptual frameworks and systems for supporting social interaction (in Japanese; translated by Masanori Sugimoto)." *Information Processing Society of Japan (IPSJ) Magazine* 40(6): 575–582. 125

Fischer, G. (2000). "Social creativity, symmetry of ignorance and meta-design." *Knowledge-Based Systems Journal* (Special Issue on Creativity & Cognition), Elsevier Science B.V., Oxford, 13(7-8): 527–537.

Fischer, G. (2001a). "Articulating the task at hand and making information relevant to it." *Human-Computer Interaction Journal*, Special Issue on "Context-Aware Computing" 16: 243–256.

Fischer, G. (2001b). "User modeling in human-computer interaction." *User Modeling and User-Adapted Interaction (UMUAI)* 11(1): 65–86. DOI: 10.1023/A:1011145532042. 5, 27, 113

Fischer, G. (2002). What's on the horizon? — Lifelong learning: New mindsets and new media. *Teachers as Scholars*, Boulder CO. 7

Fischer, G. (2003). Working and learning when the answer is not known. *The Eighth European Conference on Computer Supported Cooperative Work* Helsinki, Finland. 7

Fischer, G. (2009). End-user development and meta-design: Foundations for cultures of participation. In *End-User Development*, V. Pipek, M. B. Rossen, B. deRuyter and V. Wulf. Heidelberg, Springer: 3–14. 68

Fischer, G., E. Arias, S. Carmien, H. Eden, A. Gorman, S. i. Konomi, and J. Sullivan. (2004). "Supporting collaboration and distributed cognition in context-aware pervasive computing environments (Paper presented at the *2004 Meeting of the Human Computer Interaction Consortium "Computing Off The Desktop"*)." From http://www.cs.colorado.edu/~gerhard/papers/hcic2004.pdf. 27

Fischer, G., P. Ehn, Y. Engeström, and J. Virkkunen (2002). Symmetry of ignorance and informed participation. *Proceedings of the Participatory Design Conference (PDC'02)*, Malmö University, Sweden, CPSR.

Fischer, G. and E. Giaccardi (2006). Meta-design: A framework for the future of end user development. In *End User Develop—End User Development — Empowering people to flexibly employ advanced information and communication technology*. H. Lieberman, F. Paternò and V. Wulf, Eds. Dordrecht, The Netherlands:Kluwer Academic Publishers, pp. 427–457. 19, 107

Fischer, G. and T. Herrmann (2011). "Sociotechnical systems: A meta-design perspective." *International Journal of Sociotechnology and Knowledge Development (IJSKD)* 3(1): 1–33.

Fischer, G. and S. Konomi (2005). Innovative media in support of distributed intelligence and lifelong learning. *Proceedings of the Third IEEE International Workshop on Wireless and Mobile Technologies in Education*. Tokushima, Japan, IEEE Computer Society: 3-10. DOI: 10.1109/wmte.2005.35.

Fischer, G. and E. Scharff (2000). "Meta-design—Design for designers." *3rd International Conference on Designing Interactive Systems (DIS 2000)*, New York, ACM. DOI: 10.1145/347642.347798. 109

Fisk, A. D., Rogers, W. A., Charness, N., Czaja, S. J., and Sharit, J. (2009). *Designing for Older Adults: Principles and Creative Human Factors Approaches*, 2nd ed., Boca Raton, FL:CRC Press, J Taylor and Francis Group. DOI: 10.1201/9781420080681.

Frey, W. (2009). "Socio-technical systems in professional decision making " Retrieved June, 2009, from http://cnx.org/content/m14025/1.9/. 60, 62, 64, 65

Galvin, J. C. and C. M. Donnell (2002). Educating the consumer and caretaker on assistive technology. In *Assistive Technology: Matching Device and Consumer for Successful Rehabilitation*. M. J. Scherer, Ed. Washington, DC:American Psychological Association, pp. 153–167. DOI: 10.1037/10420-009. 5, 132

Gamma, E., R. Helm, R. Johnson, and J. Vlissides (1995). *Design Patterns—Elements of Reusable Object-Oriented Systems*. Reading, MA:Addison-Wesley Publishing Company, Inc. 10

Geels, F. W. and R. Kemp (2007). "Dynamics in socio-technical systems: Typology of change processes and contrasting case studies," *Technology in Society* 29(4): 441–455. DOI: 10.1016/j.techsoc.2007.08.009. 65

Giaccardi, E. (2005). "Metadesign as an emergent design culture." *Leonardo* 38(4): 342–349. DOI: 10.1162/0024094054762098.

Giaccardi, E. and G. Fischer (2008). "Creativity and evolution: A metadesign perspective." *Digital Creativity* 19(1): 19–32. DOI: 10.1080/14626260701847456.

Giaccardi, E. and D. Fogli (2005). "The concept of seed in meta-design: Nature, processes, and evolution." From http://x.i-dat.org/~eg/research/publications.htm.

Gorman, A. (2005). "LifeLine Website." From http://www.cs.colorado.edu/~l3d/clever/projects/lifeline.html. 34, 77, 92

Guzdial, M. (1994). "Software-realized scaffolding to facilitate programming for science learning." *Interactive Learning Environments*. 4:1 pp. 001-044. DOI: 10.1080/1049482940040101. 89

HAPTIMAP project (2012a). "HAPTIMAP website." www.haptimap.org. 106

HAPTIMAP project (2012b). "Toolkit download." http://www.haptimap.org/toolkit-download.html. 106

Hoey, J., P. Poupart, A. V. Bertoldi, T. Craig, C. Boutilier, and A. Mihailidis (2010). "Automated handwashing assistance for people with dementia using video and a partially observable Markov decision process." *Computer Vision and Image Understanding* 114: 503–519. DOI: 10.1016/j.cviu.2009.06.008. 41, 42

Hollan, J., E. Hutchins, and D. Kirsch (2001). Distributed cognition: Toward a new foundation for human-computer interaction research. In *Human-Computer Interaction in the New Millennium*. J. M. Carroll. New York:ACM Press, pp. 75–94. 26, 44, 45, 46, 49, 98

Huer, M. B. (2000). "Examining perceptions of graphic symbols across cultures: Preliminary study of the impact of culture/ethnicity." *Augmentative and Alternative Communication* (16): 180–185. DOI: 10.1080/07434610012331279034. 82

Hutchins, E. (1996). *Cognition in the Wild*. Cambridge MA:MIT Press.

IANA (Internet Assigned Numbers Authority) (1996). RFC 2045-RFC 2046. Multipurpose internet mail extensions. http://www.ietf.org/rfc/rfc2045.txt?number=2045, IANA. 114

IEEE (1990). IEEE standard glossary of software engineering terminology lEEE Std 610.121990 IEEE. http://dis.unal.edu.co/~icasta/ggs/Documentos/Normas/610-12-1990.pdf. 5

Imagine! (2007). "Imagine! website." From http://www.imaginecolorado.org. 135

IMS Global Learning Consortium (2007). "IMS learner information package accessibility for LIP information model version 1.0 final specification," from http://www.imsglobal.org/accessibility/acclipv1p0/imsacclip_infov1p0.html. 115

INCTS (2007). "V2—Information technology access interfaces." 2007, from http://www.ncits.org/tc_home/v2.htm. 115

ISO (2015). ISO/TC 173 Assistive products for people with disability. From http://www.iso.org/iso/iso_technical_committee%3Fcommid%3D53782. 3

ISO/IEC (2007a). ISO/IEC 24751-1:2008 Information technology—Individualized adaptability and accessibility in e-learning, education and training. Part 1: Framework and reference model. Geniva, Switzerland, ISO/IEC. 115

ISO/IEC (2007b). ISO/IEC 24751-2:2008 Information technology—Individualized adaptability and accessibility in e-learning, education and training Part 2: "Access for all" personal needs and preferences for digital delivery. Geniva, Switzerland, ISO/IEC.

ISO/IEC (2007c). ISO/IEC 24751-3:2008 Information technology—Individualized adaptability and accessibility in e-learning, education and training Part 3: "Access for all" digital resource description". Geniva, Switzerland, ISO/IEC.

Jakovljević, M. and L. Ostojić (2013). "Comorbidity and multimorbidity in medicine today: challenges and opportunities for bringing separated branches of medicine closer to each other." *Psychiatria Danubina* 25 (Suppl 1): 18–28. 69, 70

Juran, J. (1975). "The non-pareto principle; Mea culpa." *Quality Progress* 8: 1–3. 104

Juran, J. M. (1950). "Pareto. Lorenz, Cournot Bernoulli, Juran and others." *Industrial Quality Control 2* (October 1950): 25. 104

Karlawish, J. (2008). "Measuring decision-making capacity in cognitively impaired individuals." *Neurosignals* 16: 91–98. DOI: 10.1159/000109763. 77

King, T. (1999). *Assistive Technology—Essential Human Factors*. Boston:Allyn & Bacon, pp. 12–13. 4, 131

King, T. (2001). Ten nifty ways to make sure your clients fail with AT and AAC! (...A human factors perspective on clinical success—or not). *19th Annual Conference: Computer Technology in Special Education and Rehabilitation*. Minnespolis, MN. 4

Kintsch, A. (2002). Personal communicaiton. S. Carmien. 5, 68, 81, 134, 135

Kintsch, A. and R. dePaula (2002). A framework for the adoption of assistive technology. *SWAAAC 2002: Supporting Learning Through Assistive Technology*, Winter Park, CO:Assitive Technology Partners. 4, 5, 30, 38, 68

Knops, H. and C. Bühler (1999). *Assistive Technology on the Threshold of the New Millennium*. Amsterdam, Washington, DC, Tokyo:IOS Press.

L³D. (2006). "Center for lifelong learning and design, University of Colorado, Boulder." From http://l3d.cs.colorado.edu/. 3, 7, 8

Lancioni, G., M. O'Reilly, P. Seedhouse, F. Furniss, and C. B. (2000). "Promoting independent task performance by people with severe developmental disabilities through a new computer-aided system." *Behavior Modification* 24(5): 700–718. DOI: 10.1177/0145445500245005. 3

Lave, J. and E. Wenger (1991). *Situated Learning: Legitimate Peripheral Participation*. New York:-Cambridge University Press. DOI: 10.1017/CBO9780511815355. 23

LeCompte, M. and J. Schensul (1999). *Analyzing & Interpreting Ethnographic Data*. Walnut Creek, CA:Altamira Press. 62, 63

Lewis, C. and J. Rieman. (1993). "Task-centered user interface design: A practical introduction." From ftp://ftp.cs.colorado.edu/pub/cs/distribs/clewis/HCI-Design-Book. 5, 86

Lewis, C. H. and D. A. Norman (1986). Designing for error. In *User Centered System Design, New Perspectives on Human-Computer Interaction*. D. A. Norman and S. W. Draper, Eds. Hillsdale, NJ:Lawrence Erlbaum Associates, pp. 411–432. 26

Licklider, J. C. R. (1960). "Man-computer symbiosis." *IRE Transactions on Human Factors in Electronics* HFE-1: 4–11. DOI: 10.1109/THFE2.1960.4503259.

Lieberman, H., F. Paterno, and V. Wulf, Eds. (2006). *End User Development*. Dordrecht, The Netherlands:Kluwer Publishers. DOI: 10.1007/1-4020-5386-x. 108

LoPresti, E. F., A. Mihailidis, and N. Kirsch (2004). "Assistive technology for cognitive re-habilitation: State of the art." *Neuropsychological Rehabilitation* 14(1-2): 5–39. DOI: 10.1080/09602010343000101. 5, 23

LoPresti, E. F. B., C.; Lewis, C. (2008). "Assistive technology for cognition [Understanding the needs of people with disabilities]." *Engineering in Medicine and Biology Magazine*, IEEE 27(2): 29–39. DOI: 10.1109/EMB.2007.907396. 75

LRE for LIFE Project, U. o. T.-K., College of Education. (2001). "Steps for building instructional program packets: Selected activity analyzes (unabbreviated version)." 2003, from http://web.utk.edu/~lre4life/ftp/TADSman.PDF. 99, 100

Manguel, L. (1996). *A History of Reading*. New York:Viking. 46

Markoff, J. (2005). *What the Dormouse Said: How the Sixties Counterculture Shaped the Personal Computer Industry*, New York:Penguin Books. 15

Martin, B., and L. McCormack, (1999). Issues surrounding assistive technology use and abandon-ment in an emerging technological culture. *Proceedings of Association for the Advancement of Assistive Technology in Europe (AAATE) Conference*. Dusseldorf, Germany. 5

Martinek, V., and M. emlicka (2010). Some issues and solutions for complex navigation systems: Experience from the JRGPS project. *2010 Fifth International Conference on Systems (ICONS)* IEEE, Menuires, France: 11–16. DOI: 10.1109/ICONS.2010.24. 105

Mather, M. (2006). A review of decision making processes: Weighing the risks and benefits of aging. In *When I'm 64*. L. L. Carstensen and C.R. Hartel, Eds. Washington DC:The National Academies Press, pp. 145–173. 77

Maturana, H. R. (1997). "Metadesign." From http://www.hum.auc.dk/~rasand/Artikler/metade-sign.htm. 107

Maturana, H. R. and F. J. Varela (1987). *The Tree of Knowledge: The Biological Roots of Human Un-derstanding*. Boston, MA:Shambhala. 107

Mayer-Johnson, I. (2001). *Introduction to the Picture Communication Symbols*. Solana Beach, CA:-Mayer-Johnson, Inc. 84

Mc Sharry, J. (2014). "Challenges in managing multiple conditions: The patient experience of multimorbidity." *The European Health Psychologist* 16(6): 224–227. 3

Mihailidis, A. (2007). "Intelligent supportive environments for older adults (Coach Project)." From http://www.ot.utoronto.ca/iatsl/projects/intell_env.htm. 24, 42, 43

Mihailidis, A., J. N. Boger, T. Craig and J. Hoey (2008). "The COACH prompting system to assist older adults with dementia through handwashing: an efficacy study." *BMC Geriatrics* 8: 28. DOI: 10.1186/1471-2318-8-28. 4, 41

Mohamad, Y. (2005). Integration von Emotionaler Intelligenz in Interface-Agenten am Beispiel einer Trainingssoftware für lernbehinderte Kinder. Ph.D., Rheinisch- Westfälischen Technischen Hochschule Aachen. 54, 55, 56

Mohamad, Y., C. A. Velasco, S. Damm, and H. Tebarth (2004). Cognitive training with animated pedagogical agents (TAPA) in children with learning disabilities. In *Computers Helping People with Special Needs*. Paris, France:Springer-Verlag Berlin Heifelberg: 187–193.

Mourlas, C. and P. Germanakos (2009). *Intelligent User Interfaces: Adaptation and Personalization Systems and Technologies*. Hershey, PA:IGI. DOI: 10.4018/978-1-60566-032-5. 110, 112

Mumford, E. (1987). Sociotechnical systems design: Evolving theory and practice. *Computers and Democracy*. G. Bjerknes, P. Ehn, and M. Kyng, Eds. Aldershot, UK:Avebury, pp. 59–76. 62

Mumford, E. (2000a). A Socio-technical approach to systems design. *Requirements Engineering* 5: 59–77. DOI: 10.1007/PL00010345.

Mumford, E. (2000b). Socio-technical design: An unfulfilled promise or a future opportunity. *Proceedings of the IFIP TC9 WG9.3 International Conference on Home Oriented Informatics and Telematics*, "IF at Home: Virtual Influences on Everyday Life": Information, Technology and Society, June 2000.

Mumford, E. (2003). *Redesigning Human Systems*. Hershey, PA:Information Science Publishing. DOI: 10.4018/978-1-59140-118-6. 61

Mumford, E. (2009). "Designing human systems—The ETHICS method." Retrieved May, accessed May, 2009, from http://www.enid.u-net.com/C1book1.htm. 59, 60, 62, 64

Nardi, B. (1996a). Activity theory and human-computer interaction. *Context and Consciousness: Activity Theory and Human-Computer Interaction*. B. Nardi. Cambridge, MA:MIT Press: 7-16.

Nardi, B. (1996b). *Context and Consciousness and Consciousness Activity and Human-Computer Interactions*. Cambridge, Massachusetts, The MIT Press. 46

Nardi, B. A. (1993). *A Small Matter of Programming*. Cambridge, MA:The MIT Press. 108

Neressian, N. J., W. C. Newstetter, E. Kurz-Milcke, and J. Davies (2003). A mixed-method approach to studying distributed cognition in evolving environments. *The Fifth International Conference on Learning Sciences*. Seattle, WA.

New England ADA Center. (2015, August 2015). "ADA checklist for existing facilities." from *ADA Checklist for Existing Facilities*. www.humancentereddesign.org.

Nickerson, R. S. (1993). On the distribution of cognition: some reflections. *Distributed Cognitions Psychological and Educational Considerations*. G. Salomon. Cambridge:Cambridge University Press, pp. 229-260.

Nickerson, R. S. (2010). *Mathematical Reasoning: Patterns, Problems, Conjectures, and Proofs*. New York:Psychology Press. 79, 80

Norman, D. A. (1983). "Design rules based on analyzes of human error." *Communications of the ACM* 26(4): 254-258. 26

Norman, D. A. (1990). *The Design of Everyday Things*. New York:Currency Doubleday. 45

Norman, D. A. (1993). *Things That Make Us Smart*. Reading, MA:Addison-Wesley Publishing Company. 45, 93

O'Neill, B., K. Moran and A. Gillespie (2010). Scaffolding rehabilitation behavior using a voice-mediated assistive technology for cognition. *Neuropsychological Rehabilitation*. Vol. 20, pp. 509–527.

Olsen, R. (2000). Personal Communication. Boulder. 91, 95

Open Mobile Alliance (2007). "OMA techniclal section profile data (UAProf)." 2007, from http://www.openmobilealliance.org/tech/profiles/index.html. 114

Ostwald, J. (2003). "DynaGloss, part of the Dynasite system." 2004, from http://seed.cs.colorado.edu/dynagloss.MakeGlossaryPage.fcgi$URLinc=6. 125

Overseas Development Administration, S. D. D. (1995). "Guidance note on how to do stakeholder analysis of aid projects and programmes." London, UK. 72, 126

Patterson, D. A., G. Gibson and R. H. Katz (1988). "A case for redundant arrays of inexpensive disks (RAID)." *ACM SIGMOD Record* 17: 109–116. 24, 41

Pea, R. (1993). Practices of distributed intelligence and designs for education. In *Distributed Cognitions: Psychological and Educational Considerations*. G. Salomon, Ed. Cambridge, UK:-Cambridge University Press, pp. 47–87. 44, 45

Pea, R. D. (2004). "The social and technological dimensions of scaffolding and related theoretical concepts for learning, education, and human activity." *The Journal of the Learning Sciences* 13(3): 423–451. 87

Perkins, D. N. (1993). Person-plus: A distributed view of thinking and learning. In *Distributed Cognitions Psychological and Educational Conciderations*. G. Solomon, Ed. Cambridge. UK:Cambridge University Press, pp. 88–110.

Perkins, D. (2000). *Archimedes Bathtub: The Art and Logic of Breakthrough Thinking*. New York:W.W. Norton. 79

Perlis, A. J. (1982). Epigrams on programming. *SIGPLAN Notices*: 7–13. 108, 134

Perry, J. C., H. Zabaleta, A. Belloso and T. Keller (2009). ARMassist: A low-cost device for telerehabilitation of post-stroke arm deficits. *IFMBE Proceedings*. 25: 64–67. 95

Perry, M. (2003). Distributed cognition. In *HCI Models, Theories and Frameworks, Toward a Multidisciplinary Science*. J. M. Carroll, Ed. Burliongton, VT:Morgan Kaufman. 193–223. 45

Petroski, H. (1985). *To Engineer Is Human: The Role of Failure in Successful Design*. New York,:St. Martin's Press.

Phillips, B. and H. Zhao (1993). "Predictors of assistive technology abandonment." *Assistive Technology* 5(1). DOI: 10.1080/10400435.1993.10132205. 5, 26, 64, 68

Pollack, M. E., L. Brown, D. Colbry, C. E. McCarthy, C. Orosz, B. Peintner, S. Ramakrishnan and I. Tsamardinos (2003). "Autominder: An intelligent cognitive orthotic system for people with memory impairment." *Robotics and Autonomous Systems* 44: 273–282. DOI: 10.1016/S0921-8890(03)00077-0. 25

PTG Global (2009). "Socio-technical systems—there's more to performance than new technology." Retrieved June 10, 2009, from http://www.ptg-global.com/papers/strategy/socio-technical-systems.cfm. 60

Reason, J. (1990). *Human Error*. Cambridge, UK:Cambridge University Press. DOI: 10.1017/CBO9781139062367. 25, 26

Rehabilitation Research Design & Disability (R2D2) Center (2006). "Assistive technology outcomes measurement system project (ATOMS Project)." From http://www.uwm.edu/CHS/r2d2/atoms/. 5

Reimer-Reiss, M. (2000). Assistive technology discontinuance. *Technology and People with Disabilities Conference*. Northridge, CA. 5, 41

Reisberg, B., Steven H Ferris, Emile H Franssen, Emma Shulman, Isabel Monteiro, Steven G Sclan, Gertrude Steinberg, et al (1996). "Mortality and temporal course of probable Alzheimer's disease : A 5-Year Prospective Study." *International Psychogeriatrics* 2. DOI: 0.1017/s1041610296002657. 48

RESNA (2015). The RESNA Standards Committee on Cognitive Accessibility. http://www.resna.org/resna-standards-committee-cognitive-accessibility-ct. 3, 132

Resnick, L. B., J. M.Levine, and S. D. Teasley (1991). *Perspectives on Socially Shared Cognition*. Washington, DC:American Psychological Association. DOI: 10.1037/10096-000. 126

Ritchie, D. and K. Thompson (1974). "The UNIX Time- Sharing System." *Communications of the ACM* 17: 365-375. DOI: 10.1145/361011.361061.

Rittel, H. (1984). Second-generation design methods. In *Developments in Design Methodology*. N. Cross, Ed. New York:John Wiley & Sons, pp. 317–327.

Rittel, H. and M. M. Webber (1984). Planning problems are wicked problems. In *Developments in Design Methodology*. N. Cross, Ed. New York:John Wiley & Sons, pp. 135–144. 71, 73

Rittel, H. W. J. (1972). "On the planning crisis: Systems analysis of the first and second genera-tions." *Bedrifts Økonomen* (Norway), 8(107). 127

Rogers, Y. (1997). "An brief introduction to distributed cognition." *Cognitive Science* 24.

Rogers, Y. and J. Ellis (1994). "Distributed cognition: An alternative framework for analysing and explaining collaborative working." *Journal of Information Technology* 9: 119–128. DOI: 10.1057/jit.1994.12.

Rosson, M. B. and J. M. Carroll (2001). *Usability Engineering: Scenario-Based Development of Hu-man-Computer Interaction*. Redwood City, CA:Morgan Kaufmann. 39

Rosson, M. B., J. M. Carroll and R. K. E. Bellamy (1990). Smalltalk Scaffolding: A case study of minimalist instruction. *CHI '90*. J. C. Chew and J. Whiteside, Eds. New York:ACM, pp. 423–429.

Russell, S. and P. Norvig (2009). *Artificial Intelligence: A Modern Approach* 3rd ed., Upper Saddle River, NJ:Peterson. 4, 97

Salomon, G., Ed. (1993). *Distributed Cognitions: Psychological and Educational Considerations*. Cam-bridge, UK:Cambridge University Press. 44

Saskatchewan Learning—Special Education Unit. (2003). "Task Analysis." From http://www.sasked.gov.sk.ca/k/pecs/se/docs/meeting/s6analysis.html. 98

Sawyer, R. K. (2006). "The Cambridge handbook of the learning sciences." *Learning* 190: xix, 627. 51

Scacchi, W. (2004). Socio-Technical Design. The Encyclopedia of Human-Computer Interac-tion. From https://www.interaction-design.org/literature/book/the-encyclopedia-of-hu-man-computer-interaction-2nd-ed.

Schank, R. C. and R. P. Abelson (1975). Scripts, plans, and knowledge, Yale University. 97

Schank, R. C. and R. P. Abelson (1977). *Scripts, Plans, Goals, and Understanding*. Hillsdale, NJ:Law-rence Erlbaum Associates, Inc. 49, 98, 101

Scherer, M. (2011). *Assistive Technologies and Other Supports for People With Brain Impairment*, New York:Springer Publishing Company. 3, 133

Scherer, M. J. (1996). *Living in the State of Stuck: How Technology Impacts the Lives of People with Disabilities*. Cambridge:Brookline Books. 5, 137

Scherer, M. J. and J. C. Galvin (1996). An outcomes perspective of quality pathways to the most appropriate technology. *Evaluating, Selecting and Using Appropriate Assistive Technology*. M. J. Scherer and J. C. Galvin. Gaithersburg, MD:Aspen Publishers, Inc, pp. 1–26. 5

Scherer, M. J., Galvin, J.C. (1994). "Matching people with technology." *Rehab Management* 7(2): 128–130.

Scott, J. (2000). *Social Network Analysis: A Handbook*, 2nd Ed. Newberry Park, CA:Sage. 61

Simon, H. A. (1984). The structure of ill-structured problems. In *Developments in Design Methodology*. N. Cross, Ed. New York:John Wiley & Sons, pp. 145–166. 71

Simon, H. A. (1996). *The Sciences of the Artificial*. Cambridge, MA:The MIT Press. 73, 79, 81, 125

Snell, M. E. (1987). *Systematic Instruction of People with Severe Handicaps*. Columbus, OH:Merrill Publishing Company. 82, 93

Solomon, G. (1993). *Distributed Cognitions Psychological and Educational Conciderations*. Cambridge, UK:Cambridge University Press. 98

Sommerville, I. (2001). *Software Engineering*. Reading, MA:Addison-Wesley. 24

Stephanidis, C. and A. ASavidis (2001). "Universal access in the information society: Methods, tools, and interaction technology." *Universal Access in the Information Society* 1(1): 40–55. 6

Suchman, L. (1987). *Plans and Situated Actions: The Problem of Human-Machine Communication*. Cambridge, UK:Cambridge University Press. 3, 23, 26, 45, 57, 58, 64

Suchman, L. A. (2007). *Human-Machine Recongfigurations, Plans and Situated Actions*, 2nd ed. New York:Cambridge University Press. 95

Sullivan, J. (2004). "Mobility for All project." 2004, from http://www.cs.colorado.edu/~l3d/clever/projects/mobility.html. 23, 113

Sullivan, J. F. (2003). Lost on the front range bus systems. C. project. L3D, Boulder CO. 23, 33

Sullivan, J. F. (2005). "Mobility for All." From http://l3d.cs.colorado.edu/clever/projects/mobility.html. 92

Sumner, T. (1995). The high-tech toolbelt: A study of designers in the workplace. In *Proceedings of ACM CHI'95 Conference on Human Factors in Computing Systems*. I. Katz, R. Mack and L. Marks, Eds. New York:ACM. 1: 178-185. DOI: 10.1145/223904.223927. 6

Taylor, F. (1967). *Principles of Scientific Management*. New York (first published 1911):Norton and Co. 59

Tebarth, H., Y. Mohamad, and M. Pieper (2000). "Cognitive training by animated pedagogical agents (TAPA) development of a tele-medical system for memory improvement in children with epilepsy." *Workshop Proceedings of the 6th ERCIM Workshop*, Florence, Italy, pp. 158–168. 54

Technologies, A. "Visual Impact memory and skill prompter." From http://www.ablelinktech.com.

trace.wisc.edu (2005). "V2 support project at Trace—toward a universal remote console standard." 2007, from http://trace.wisc.edu/urc/. 114, 125

URC Consortium (2007). "The universal remote console consortium." from http://myurc.org/. 114

Vanderheiden, G. (1992). A brief look at technology and mental retardation in the 21st century. *Mental Retardation in the Year 2000*. L. Rowitz, Ed. New York:Springer-Verlag, Inc., pp. 268–278.

Velasco, C. A., Y. Mohamad, D. Stegemann, H. Gappa, G. Nordbrock, E. Hartsuiker, J. Sánchez-Lacuesta, and J. M. Belda (2004). "IPCA: Adaptative interfaces based upon biofeedback sensors" *Computers Helping People with Special Needs, Lecture Notes in Computer Science*, 3118/2004: pp. 129–134. 54

Voltaire; Peter Gay (1962). Philosophical Dictionary (Dictionnaire philosophique, 1774). New York: Basic Books, Inc. 102

VTT (2013). Slide from developers meeting. S. V. Heinonen. Helsinki, VTT. 37, 106

W3C (2007). "Composite Capability/Preference Profiles (CC/PP): Structure and Vocabularies 2.0." From http://www.w3.org/TR/2007/WD-CCPP-struct-vocab2-20070430/.

W3C (2009a). "WAI-ARIA overview." Retrieved December 14, 2009, from http://www.w3.org/WAI/intro/aria. 114, 115

W3C (2009b). "Web Accessibility Initiative (WAI)." From http://www.w3.org/WAI/. 124

Weimann, F., J. Seybold and B. Hofmann-Wellenhof (2007). Development of a pedestrian navigation system for urban and indoor environments. *Proceedings of the 20th International Technical Meeting of the Satellite Division of The Institute of Navigation (ION GNSS 2007)*, Fort Worth, TX. 37

Weinberger, A., K. Stegmann and F. Fischer (2005). Computer-supported collaborative learning in higher education: Scripts for argumentative knowledge construction in distributed groups. *The Next 10 Years! Proceedings of the 2005 Conference on Computer Support for Collaborative Learning, CSCL '05*, pp. 717–726. DOI: 10.3115/1149293.1149387. 97

Wells, H. G. (1905). *The Time Machine: An Invention*. London:William Heinemann. 90

Wenger, E. (1998). *Communities of Practice — Learning, Meaning, and Identity*. Cambridge, UK:-Cambridge University Press. DOI: 10.1017/CBO9780511803932. 126

Wilkinson, K. M. and W. J. McIlvane (2002). Considerations in teaching graphic sysmbols to beginning communicators. In *Exemplary Practics of Beginning Communicators*. J. Reichle, D. R. Beukelman and J. C. Light, Eds. Baltiimore MD:Paul H.Brookes Publishing, pp. 273–321. 82

Wireless Application Protocol Forum, L. (2001). "WAG UAProf Version 20." From http://www.openmobilealliance.org/tech/affiliates/wap/wap-248-uaprof-20011020-a.pdf. 120

Wood, D., J. S. Bruner and G. Ross (1976). "The role of tutoring in problem-solving." *Journal of Child Psychology and Psychiatry* 17: 89–100. DOI: 10.1111/j.1469-7610.1976.tb00381.x. 52, 89

Wood, D. and D. Middleton (1975). "A study of assisted problem-solving." *British Journal of Psychology* 66: 181–191. DOI: 10.1111/j.2044-8295.1975.tb01454.x. 52

World Health Organization (2001). International Classification of Functioning, Disability and Health (ICF), World Health Organization (WHO). 4, 132

Yates, F. A. (1966). *The Art of Memory*, Chicago, IL:University of Chicago Press. 81

Zimmermann, G., G. Vanderheiden, and C. Rich. (2006). "Universal control hub & task-based user interfaces." 2007, from http://myurc.org/publications/2006-Univ-Ctrl-Hub.php. 125

Author Biography

Dr. Stefan Carmien is a staff scientist at the Tecnalia foundation. His work focuses on ubiquitous and mobile assistive technologies and context-aware systems, involving the study of the socio-technological environment, its context, and the human user and deep personal configuration (meta-design) and end-user programming (within a distributed cognition framework) as a solution to technology abandonment. He has also performed studies on image recognition and cognitive disabilities in memory and recall. He was most recently the PI for ASSISTANT, a 2.4 M € EC project supporting use of public transportation by elders with disabilities. Stefan holds a Ph.D. in computer science with a certificate in cognitive science from the University of Colorado. His Ph.D. work centered on the design of systems for active task support for people with cognitive disabilities and caregivers. He has contributed many peer-reviewed articles, six book chapters, and is the author of the book *Leveraging Skills into Independent Living – Distributed Cognition and Cognitive Disability.*

Printed in the United States
by Baker & Taylor Publisher Services